New Atlantis
and
The Great Instauration

Crofts Classics Series

FRANCIS BACON

New Atlantis and The Great Instauration

Second Edition

Edited by

Jerry Weinberger

WILEY Blackwell

This second edition first published 2017
© 2017 John Wiley & Sons, Inc

Edition History: Harlan Davidson, Inc. (1e, 1980; revised edition, 1989)

Registered Office
John Wiley & Sons Ltd, The Atrium, Southern Gate, Chichester, West Sussex, PO19 8SQ, UK

Editorial Offices
350 Main Street, Malden, MA 02148-5020, USA
9600 Garsington Road, Oxford, OX4 2DQ, UK
The Atrium, Southern Gate, Chichester, West Sussex, PO19 8SQ, UK

For details of our global editorial offices, for customer services, and for information about how to apply for permission to reuse the copyright material in this book please see our website at www.wiley.com/wiley-blackwell.

The right of Jerry Weinberger to be identified as the author of the editorial material in this work has been asserted in accordance with the UK Copyright, Designs and Patents Act 1988.

Library of Congress Cataloging-in-Publication Data

Names: Bacon, Francis, 1561-1626, author. | Weinberger, Jerry, 1944- editor.
 | Bacon, Francis, 1561-1626. New Atlantis. | Bacon, Francis, 1561-1626.
 Instauratio magna. English.
Title: New Atlantis ; and, The great instauration / edited by Jerry
 Weinberger.
Description: Second Edition. | Hoboken : Wiley, 2016. | Includes
 bibliographical references.
Identifiers: LCCN 2015049558| ISBN 9781119097983 (cloth) |
 ISBN 9781119098027 (pbk.)
Subjects: LCSH: Philosophy–Early works to 1800. | Philosophy of
 nature–Early works to 1800. | Science–Philosophy–Early works to 1800. |
 Utopias–Early works to 1800.
Classification: LCC B1155 2016 | DDC 100–dc23 LC record available at
 http://lccn.loc.gov/2015049558

A catalogue record for this book is available from the British Library.

Cover image: Portrait of Francis Bacon by Paul van Somer, Di Agostini Picture Library/ Getty Images

Set in 10/12 pt SabonLTStd-Roman by Thomson Digital, Noida, India
Printed and bound in Malaysia by Vivar Printing Sdn Bhd

1 2017

Contents

Preface: Why a Second Edition?

The Crofts Classics edition of Sir Francis Bacon's *New Atlantis and The Great Instauration* has been in classroom use since the late 1970s, having been revised to correct typography in 1980. Since those early years much scholarship, including some of my own, has been published on Bacon in general and on these two texts. I therefore thought a significantly revised second edition was in order, especially since some of my earlier conclusions about these works of Bacon had changed, even if only slightly.

Given the opportunity to bring a new edition into print with Wiley, I thought it best to increase the amount of material in order to provide student-readers wider context in which to consider Bacon's project as a whole and *New Atlantis* in particular. To this end I decided first to provide the aphorisms on the Idols of the Mind from *The Great Instauration: Novum organum Book One*. The purpose of this addition is to make clear how much Bacon thought the problems of human reason resulted from reason's inherent weaknesses, not just the weight of obscure traditions. In order to broaden the political context for *New Atlantis* I have included two of Bacon's *Essays*: "Of Unity in Religion" and "Of the True Greatness of Kingdoms and Estates." These two essays, along with reference in the interpretive essay to Bacon's 1622 *The History of the Reign of King Henry the Seventh*, link Bacon's utopian fable to some of his most concrete considerations of practical political life, especially as regards religion and war. I decided as well to place the interpretive essay on *New Atlantis* and Bacon's essays at the end of the volume, as it will be more useful to students *after* they have read Bacon's texts.

Acknowledgments

For his warm and helpful support for this second edition I would like to thank Andrew J. Davidson, Senior Editor (History, Social Sciences & Humanities) at John Wiley & Sons, Inc. The project could not have gone forward without the indispensable help of Project Editor Julia Kirk and Editorial Assistant Maddie Koufogazos. Project Manager Ian Critchley did a masterful job overseeing the copy editing, typesetting, and proofreading stages of production. My wife Diane, as usual whenever I'm absorbed in a book project, put up with much neglect while I worked on this new edition. Her cheerfulness during my distraction never fails to amaze.

Note on the Texts

The text of the *New Atlantis* is reprinted from the widely respected, once standard edition of Bacon's works by Spedding, Ellis, and Heath. The original spelling and punctuation have everywhere been retained. The *New Atlantis* was first published by Rawley in 1627, after Bacon's death, and no original manuscript of the work remains. All the editions of the *New Atlantis* since 1627 have been reprints of Rawley's first edition, and fortunately there are no serious variations in the editions. The 1627 edition has "Solamona" and "Salomon's House," while the 1658 and 1670 editions, among others, have "Salomona" and "Solomon's House." Spedding follows the 1627 edition and is supported in this choice by the Latin translation. Although Spedding is probably correct, the variation makes no difference at all in the meaning of the text. The 1627 and 1670 editions have "It so fell out, that there was in one of the boats, one of our wise men of the society of Salomon's House . . . " while the 1658 edition, which Spedding follows, has "It so fell out, that there was in one of the boats one of the wise men of the society of Solomon's House . . . " (see p. 74, below). In this instance, Spedding is perhaps wrong, since the 1627 and 1670 texts are supported by the Latin translation. But Spedding's text is not emended in this volume, because, again, the variation makes no difference at all in the meaning of the text. The 1627 edition and Spedding have "came aboard" (see p. 64, below), while the 1658 edition has "made aboard"; and the 1627 edition has "that we knew he spake it . . . " (see p. 77, below), while the 1658 edition and Spedding have "that we knew that he spake it. . . . " These minor variations require no emendation. No critical problems in the text have so far been discovered that would warrant rejecting Spedding's text as the standard edition of the *New Atlantis. The Great Instauration* was published originally in Latin in 1620 as *Instauratio magna*. The translation used in this volume for the preliminaries and the aphorisms of *Novum organum* on the Idols of the Mind is Spedding's widely reprinted translation. Although Spedding's translation was long considered to be the standard, it is sometimes loose, and two important corrections have been noted (see pp. 21, 32, below). Volume XI of the *Oxford Francis Bacon*, edited by the late Graham Rees, will doubtless become the standard and, it is to be hoped, widely reprinted translation and edition. I have referred to it as necessary for this classroom edition.

Principal Dates in the Life of Sir Francis Bacon

1561	Born in London, son of Lord Keeper of the Seal.
1573–1575	Studied at Cambridge.
1576	Enrolled at Gray's Inn, where he studied for less than one year.
1582	Became a barrister.
1584	Sat in the Commons. Bacon was an influential member of the House of Commons for the next thirty-six years.
1597	*Essays*.
1603	Knighted by James I.
1605	*Advancement of Learning*.
1612	*Essays*, enlarged.
1613	Appointed Attorney General.
1618	Made Lord Chancellor; created Baron Verulam.
1620	*The Great Instauration: Novum organum*.
1621	Created Viscount St. Albans; impeached for accepting bribes.
1622	*The History of the Reign of King Henry the Seventh*.
1624	Most likely date by which Bacon had written *New Atlantis*.
1625	*Essays*, greatly enlarged.
1626	Died April 9.
1627	*New Atlantis* published.

Introduction to the Second Edition

Along with Machiavelli, Hobbes, and Descartes, Francis Bacon was one of the founders of modern thought. These thinkers coupled realistic politics with a new science of nature in order to transform the age-old view of humanity's place in the world. They contended that once the efforts of the human intellect were directed from traditional concerns to new ones—from contemplation to action, from the account of what people ought to do to what they actually do, and from metaphysics to the scientific method for discovering natural causes—the harsh inconveniences of nature and political life would be relieved and overcome. No longer to be revered or endured, the realms of nature and society would become the objects of human control.

Bacon called his version of this project the "Great Instauration," an ambiguous term that means at once great restoration and great founding. But he left no doubt that he was engaged in something altogether new: His restoration—his reform of the ways and means of human reason—would in fact be a founding because its aim would be "to lay the foundation, not of any sect or doctrine, but of human utility and power," in order to "command nature in action."[1] Bacon did not, however, think his instauration would be quick or easy. It would surely provoke opposition from political, theological, and academic interests with something to lose in a new intellectual order. But for Bacon, the larger problem was that the impediments to the power of reason are deeply rooted in the character of reason itself. In his famous doctrine of the "Idols of the Mind," he outlined four categories of defects that infect and mislead human reason.

First, reason seeks more order in the world than actually exists and it gets fooled into thinking that impossible things

[1]*Below*, pp. 17, 21.

exist. Second, reason tends to become obsessed with one cause that it thinks explains everything. Third, words fool reason, as if all words refer to something real. Finally, reason has been seduced by a long history of fruitless and quarrelsome philosophical speculation. Bacon has an especial dislike for the baleful influence, still active at his time, of Plato and Aristotle. Their dogmatic preference for contemplation over action reflected contempt for the practical arts, a contempt much more harmful than noble. For it merely served to hide the real courses of nature from view, so that from Aristotle one hears "the voice of dialectics more often than the voice of nature" and in Plato one sees that "he infected and corrupted natural studies by his theology as much as Aristotle did by his dialectic."[2]

For Bacon, these defects of reason had first to be exposed and rooted out as far as possible before reason's true powers could be unleashed. Then, an entirely new science, based on careful induction and especially experimentation, could discover the latent actions of nature and bend them to human purposes. Human life needs tools for action, but from the ancient wisdom we got nothing but theological and metaphysical claptrap. Moreover, the ancients applied this intellectual junk to practical and political affairs, especially after Socrates, who was famous for having brought philosophy down from the heavens. But the ancients' concern for practical affairs was in fact impractical and served merely to fuel violent controversies about justice and the best regime, controversies that are inevitable when we are faced by material scarcity and a cosmos that is hostile to our wills and indifferent to our needs.

For Bacon, reason directed by new means and ends would endow human beings with powers over nature unimaginable by his contemporaries. He wasn't restrained in what he predicted. Bacon's new science would put nature on the rack: it gives up its secrets "when by art and the hand of man she is

---

[2]"*The* Refutation of Philosophies" in *The Philosophy of Francis Bacon*, ed. Benjamin Farrington (Liverpool: Liverpool University Press, 1964), pp. 112–15.

forced out of her natural state, and squeezed and moulded."[3] When the latent actions and structures of nature are then genuinely understood, he argued, the dreams of the alchemists really will come true: it will be possible to transform something not gold into the real McCoy.[4] And so it is: with atomic science it is possible today to turn mercury into gold, and that's no different in principle from turning oil into polyester cloth for a suit of clothes. Bacon did not blush at taking aim even at the corruptibility and mortality of the human body, the conquest of which he called the "noblest work" of natural philosophy.[5] Bacon's project was thus far more than a mere reformation of human reason and even more than the founding of a new intellectual or political order. For every founding prior to his own had been limited by something beyond the powers of the founder, whether it be nature, fortune, or God. With Bacon's project we encounter an altogether new horizon, one that forces a reconsideration of the words: "In the beginning . . ."

The Great Instauration and New Atlantis

It is fashionable now to have doubts about modern science and technology and to acknowledge that their blessings can be mixed. But even so, we entertain such thoughts from within a world already transformed, mostly for the better, by them: In our everyday lives we interact constantly with their products to the point of taking them for granted. Imagine if you will, what life must have been like before the discovery of anaesthesia or when women had lots of babies because so many newborns died. In making the case for his Great Instauration, Bacon

[3] *Below*, p. 28.
[4] *Novum organum*, 2: 1–5, in *The Works of Francis Bacon*, ed. James Spedding, Robert Leslie Ellis, and Douglas Denon Heath, 14 vols. (London: Longman and Co., etc. 1857–74), hereafter BW, IV, 119–123.
[5] *De sapientia veterum* XI, BW, VI, 646, 721; below, pp. 21, 31–32.

could only conjecture about the future powers of modern science and technology. A quick look at the plan laid out in his 1620 *The Great Instauration* shows that Bacon knew perfectly well that his project would be a matter of generations and far beyond his powers to effect. The plan describes six parts, most of which are vastly incomplete. The first part was to be a survey of the knowledge presently available and of forms of knowledge that have so far been omitted and need to be developed. Bacon tells us that this first part is "wanting," but that "some account" of the matter can be found in the second book of his (1605) *Proficience and Advancement of Learning, Divine and Human.* The second part was to present the art of interpreting nature, by which Bacon means the discovery of the latent motions and structures of natural phenomena. That part is *Novum organum*, which follows *The Great Instauration* in the 1620 volume. But it is but a "summary digested into aphorisms" and is in fact largely incomplete.[6] The third part was to be a natural history cataloging all the phenomena of nature. In the 1620 volume, Bacon provides not that history but rather a list of one hundred and thirty different histories to be written on natural phenomena such as winds, comets, seas, gems and stones, the human heart and pulse, and so on (although he did write the history of winds, a history of life and death, and a history of dense and rare). The fourth part was to be a demonstration of how the new art of interpreting nature would be applied to those natural histories, but of this part we have but a few fragments. The fifth part would be a temporary collection of conclusions derived by Bacon from his ordinary reasoning, later to be tested by the new art of interpreting nature. Again, we have just a preface. The sixth part would consist of the complete philosophy set out by the plan. Of this, of course, nothing exists and Bacon himself comments

---

[6]*See* the late Graham Rees's excellent and meticulous introduction to The *Instauratio magna* Part II: *Novum organum* and Associated Texts (with Maria Wakely) in *The Oxford Francis Bacon*, Vol. XI (Oxford University Press, 2004), hereafter Rees.

that it is a "thing both above my strength and beyond my hopes."[7]

So what can we learn about our time from Sir Francis Bacon? Not all that much about natural science and technology as we know them. But that said, Bacon thought seriously about what it would *mean* for humankind to possess the power over nature he envisioned and we now have. He knew that the scientific transformation of the world would have extraordinary moral and political consequences and that it would pose new problems in place of old ones: In a world to be conquered rather than endured, what moral and political principles will guide the human energies released by the activity of conquest? And in a world freed from God's providence—a world in which we literally choose our own destiny—will anything limit the objects of our choice? And what are the best means for overcoming the resistance of those who would stand to lose with the passing of the old world, or of those too timid to welcome wholeheartedly a new and unknown one? Bacon's teaching about modern science is ultimately about human relations during and after the conquest of nature. Thus the core of his teaching about science is to be found not so much in his account of scientific investigation, but rather in his moral and political thought.

Of part six of his plan, Bacon says that if it were to exist, it would describe the ends of science, where "human Knowledge and human Power, do really meet in one"; and of such a picture Bacon says not that it is beyond his powers and so impossible for him to provide but rather that, given "the present condition of things and men's minds," it "cannot easily be conceived or imagined."[8] This means, of course, that with difficulty such a picture can be conceived or imagined and is not beyond Bacon's strength and hope. By 1624 Bacon had written in his *New Atlantis* just such a picture of a society formed by and dedicated to natural science and the technological conquest of nature. Bacon names this society "Bensalem," which means

[7]*Below*, p. 31.
[8]*Ibid.*

"offspring of peace, perfection, or wholeness." Bensalem is, moreover, dedicated to ends that are "compounded of all goodness."[9]

Bacon, however, did not publish *New Atlantis* during his lifetime and it did not see the light of day until 1627. By long tradition *New Atlantis* has been thought to be incomplete. According to Rawley, Bacon's secretary who oversaw its publication, the work was left unfinished because Bacon was deterred by considerations of time and preference from composing the "frame of Laws, or of the best state or mould of a commonwealth," an account of scientific *politics*, that would have completed it.[10] But before we accept Rawley's opinion, it is important to remember Bacon's claim that the sixth part of his project would be difficult to conceive or imagine. In *The Great Instauration* Bacon explains this difficulty as the result of men thinking "the matter in hand" to be "mere felicity of speculation" rather than the mastery of nature.[11] But in his division of the sciences in the *Advancement of Learning*, Bacon provides an additional explanation for this difficulty, at least as regards politics. There he says that government is a "part of knowledge secret and retired, in both these respects in which things are deemed secret; for some things are deemed secret because they are hard to know, and some because they are not fit to utter."[12] If the end of science is difficult to imagine, this may well be because the account of its politics must be "secret and retired." Moreover, in his 1623 Latin version (*De augmentis scientiarum*) and expansion of the *Advancement of Learning*, Bacon comments that if his "leisure time shall hereafter produce anything concerning political knowledge, the

---

[9]*Below*, pp. 72, 87. For a fuller account of the place of *New Atlantis* in the whole of Bacon's corpus, see Jerry Weinberger, "Science and Rule in Bacon's Utopia: An Introduction to the Reading of *New Atlantis*," *American Political Science Review* (1976), pp. 866–72.
[10]*Below*, pp. 62, 111.
[11]*Below*, p. 31.
[12]*BW*, III, 473–76.

work will perchance be either abortive or posthumous."[13] This describes *New Atlantis* to a T. It is entirely possible, then, that *New Atlantis* may only *appear* to be incomplete and may indeed be the picture of the end of science as anticipated in the sixth part of Bacon's plan: since it's about the politics of science its teaching is "secret and retired" and to be discerned only with difficulty. We may well have to read very carefully and between Bacon's posthumous lines. Moreover, if we take Bacon at his word, something about the politics of modern science is not just hard to know but also not fit to utter. Just what might that be?

---

[13]*De augmentis* BW, V, 79.

THE
GREAT INSTAURATION

PROŒMIUM[1]

FRANCIS OF VERULAM[2]

REASONED THUS WITH HIMSELF,

AND JUDGED IT TO BE FOR THE INTEREST OF THE PRESENT AND FUTURE GENERATIONS THAT THEY SHOULD BE MADE ACQUAINTED WITH HIS THOUGHTS.

Being convinced that the human intellect makes its own difficulties, not using the true helps which are at man's disposal soberly and judiciously; whence follows manifold ignorance of things, and by reason of that ignorance mischiefs innumerable; he[3] thought all trial should be made, whether that commerce[4] between the mind of man and the nature of things, which is more precious than anything on earth, or at least than anything that is of the earth, might by any means be restored to its perfect and original condition, or if that may not be, yet reduced[5] to a better condition than that in which it now is. Now that the errors which have hitherto prevailed, and which will prevail for ever, should (if the mind be left to go its own

[1]*Preliminary comment, prelude.*
[2]*Bacon* was Baron of Verulam.
[3]*Throughout* the Proœmium Bacon refers to himself.
[4]*Intercourse.*
[5]*Brought* back.

New Atlantis and The Great Instauration, Second Edition.
Edited by Jerry Weinberger.
© 2017 John Wiley & Sons, Inc. Published 2017 by John Wiley & Sons, Inc.

way), either by the natural force of the understanding or by help of the aids and instruments of Logic, one by one correct themselves, was a thing not to be hoped for: because the primary notions of things which the mind readily and passively imbibes, stores up, and accumulates (and it is from them that all the rest flow) are false, confused, and overhastily abstracted from the facts; nor are the secondary and subsequent notions less arbitrary and inconstant; whence it follows that the entire fabric of human reason which we employ in the inquisition of nature, is badly put together and built up, and like some magnificent structure without any foundation. For while men are occupied in admiring and applauding the false powers of the mind, they pass by and throw away those true powers, which, if it be supplied with the proper aids and can itself be content to wait upon nature instead of vainly affecting to overrule her, are within its reach. There was but one course left, therefore,—to try the whole thing anew upon a better plan, and to commence a total reconstruction of sciences, arts, and all human knowledge, raised upon the proper foundations. And this, though in the project and undertaking it may seem a thing infinite and beyond the powers of man, yet when it comes to be dealt with it will be found sound and sober, more so than what has been done hitherto. For of this there is some issue;[6] whereas in what is now done in the matter of science there is only a whirling round about, and perpetual agitation, ending where it began. And although he was well aware how solitary an enterprise it is, and how hard a thing to win faith and credit for, nevertheless he was resolved not to abandon either it or himself; nor to be deterred from trying and entering upon that one path which is alone open to the human mind. For better it is to make a beginning of that which may lead to something, than to engage in a perpetual struggle and pursuit in courses which have no exit. And certainly the two ways of contemplation are much like those two ways of action, so much celebrated, in this—that the one, arduous and difficult in the beginning, leads out at last into the open country; while the other, seeming at

[6]*He* has already produced some results.

first sight easy and free from obstruction, leads to pathless and precipitous places.

Moreover, because he knew not how long it might be before these things would occur to any one else, judging especially from this, that he has found no man hitherto who has applied his mind to the like, he resolved to publish at once so much as he has been able to complete. The cause of which haste was not ambition for himself, but solicitude for the work; that in case of his death there might remain some outline and project of that which he had conceived, and some evidence likewise of his honest mind and inclination towards the benefit of the human race. Certain it is that all other ambition whatsoever seemed poor in his eyes compared with the work which he had in hand; seeing that the matter at issue is either nothing, or a thing so great that it may well be content with its own merit, without seeking other recompence.

EPISTLE DEDICATORY[7]

TO OUR MOST GRACIOUS AND MIGHTY
PRINCE AND LORD

JAMES,[8]
BY THE GRACE OF GOD
OF GREAT BRITAIN,
FRANCE, AND IRELAND
KING,

DEFENDER OF THE FAITH, ETC.

Most Gracious and Mighty King,

Your Majesty may perhaps accuse me of larceny, having stolen from your affairs so much time as was required for this work. I know not what to say for myself. For of time there can be no restitution, unless it be that what has been abstracted from your business may perhaps go to the memory of your name and the honour of your age; if these things are indeed worth anything. Certainly they are quite new; totally new in their very kind: and yet they are copied from a very ancient model; even the world itself and the nature of

---

[7]*Letter* of dedication.
[8]*James* I, King of England, 1603–1625

New Atlantis and The Great Instauration, Second Edition.
Edited by Jerry Weinberger.
© 2017 John Wiley & Sons, Inc. Published 2017 by John Wiley & Sons, Inc.

things and of the mind. And to say truth, I am wont for my own part to regard this work as a child of time rather than of wit;[9] the only wonder being that the first notion of the thing, and such great suspicions concerning matters long established, should have come into any man's mind. All the rest follows readily enough. And no doubt there is something of accident (as we call it) and luck as well in what men think as in what they do or say. But for this accident which I speak of, I wish that if there be any good in what I have to offer, it may be ascribed to the infinite mercy and goodness of God, and to the felicity of your Majesty's times; to which as I have been an honest and affectionate servant in my life, so after my death I may yet perhaps, through the kindling of this new light in the darkness of philosophy, be the means of making this age famous to posterity; and surely to the times of the wisest and most learned of kings belongs of right the regeneration and restoration of the sciences. Lastly, I have a request to make—a request no way unworthy of your Majesty, and which especially concerns the work in hand; namely, that you who resemble Solomon in so many things—in the gravity of your judgments, in the peacefulness of your reign, in the largeness of your heart, in the noble variety of the books which you have composed—would further follow his example[10] in taking order[11] for the collecting and perfecting of a Natural and Experimental History, true and severe[12] (unincumbered with literature and book-learning), such as philosophy may be built upon,—such, in fact, as I shall in its proper place describe: that so at length, after the lapse of so many ages, philosophy and the sciences may no longer float in air, but rest on the solid foundation of

[9]*A* product of favorable circumstances rather than of Bacon's intellect.
[10]*See* I Kings 4: 29–33.
[11]*To* command that it be done.
[12]*Serious*, accurate.

experience of every kind, and the same well examined and weighed. I have provided the machine, but the stuff must be gathered from the facts of nature. May God Almighty long preserve your Majesty!

> *Your Majesty's*
> > *Most bounden and devoted*
> > > *Servant,*
>
> # FRANCIS VERULAM,
> ### Chancellor

THE GREAT INSTAURATION[13]

PREFACE

That the state of knowledge is not prosperous nor greatly advancing; and that a way must be opened for the human understanding entirely different from any hitherto known, and other helps provided, in order that the mind may exercise over the nature of things the authority which properly belongs to it.

It seems to me that men do not rightly understand either their store[14] or their strength, but overrate the one and underrate the other. Hence it follows, that either from an extravagant estimate of the value of the arts which they possess, they seek no further; or else from too mean an estimate of their own powers, they spend their strength in small matters and never put it fairly to the trial in those which go to the main.[15] These are as the pillars of fate set in the path of knowledge;[16] for men have neither desire nor hope to encourage them to penetrate further. And since opinion of store[17] is one of the chief causes of want, and satisfaction with the present induces neglect of provision for the future, it becomes a thing not only useful, but absolutely necessary, that the excess of honour and admiration with which our existing stock of inventions is regarded be in the very entrance and threshold of the work, and that frankly and

[13]*Restoration* after decay; setting up or founding.
[14]*Tools.*
[15]*The* most important matters.
[16]*These* defects act as the limits of human effort.
[17]*Plenty.*

New Atlantis and The Great Instauration, Second Edition.
Edited by Jerry Weinberger.
© 2017 John Wiley & Sons, Inc. Published 2017 by John Wiley & Sons, Inc.

without circumlocution, stripped off, and men be duly warned not to exaggerate or make too much of them. For let a man look carefully into all that variety of books with which the arts and sciences abound, he will find everywhere endless repetitions of the same thing, varying in the method of treatment, but not new in substance, insomuch that the whole stock, numerous as it appears at first view, proves on examination to be but scanty. And for its value and utility it must be plainly avowed that that wisdom which we have derived principally from the Greeks is but like the boyhood of knowledge, and has the characteristic property of boys: it can talk, but it cannot generate; for it is fruitful of controversies but barren of works. So that the state of learning as it now is appears to be represented to the life in the old fable of Scylla, who had the head and face of a virgin, but her womb was hung round with barking monsters, from which she could not be delivered. For in like manner the sciences to which we are accustomed have certain general positions which are specious and flattering; but as soon as they come to particulars, which are as the parts of generation, when they should produce fruit and works, then arise contentions and barking disputations, which are the end of the matter and all the issue they can yield. Observe also, that if sciences of this kind had any life in them, that could never have come to pass which has been the case now for many ages—that they stand almost at a stay,[18] without receiving any augmentations worthy of the human race; insomuch that many times not only what was asserted once is asserted still, but what was a question once is a question still, and instead of being resolved by discussion is only fixed and fed; and all the tradition and succession of schools is still a succession of masters and scholars, not of inventors and those who bring to further perfection the things invented. In the mechanical arts we do not find it so; they, on the contrary, as having in them some breath of life, are continually growing and becoming more perfect. As originally invented they are commonly rude, clumsy, and shapeless; afterwards they acquire new powers

[18]*Stop.*

and more commodious arrangements and constructions; in so far that men shall sooner leave the study and pursuit of them and turn to something else, than they arrive at the ultimate perfection of which they are capable.[19] Philosophy and the intellectual sciences, on the contrary, stand like statues, worshipped and celebrated, but not moved or advanced. Nay, they sometimes flourish most in the hands of the first author, and afterwards degenerate. For when men have once made over their judgments to others' keeping, and (like those senators whom they called *Pedarii*)[20] have agreed to support some one person's opinion, from that time they make no enlargement of the sciences themselves, but fall to the servile office of embellishing certain individual authors and increasing their retinue. And let it not be said that the sciences have been growing gradually till they have at last reached their full stature, and so (their course being completed) have settled in the works of a few writers; and that there being now no room for the invention of better, all that remains is to embellish and cultivate those things which have been invented already. Would it were so! But the truth is that this appropriating[21] of the sciences has its origin in nothing better than the confidence of a few persons and the sloth and indolence of the rest. For after the sciences had been in several parts perhaps cultivated and handled diligently, there has risen up some man of bold disposition, and famous for methods and short ways which people like, who has in appearance reduced them to an art, while he has in fact only spoiled all that the others had done. And yet this is what posterity like, because it makes the work short and easy, and saves further inquiry, of which they are weary and impatient. And if any one take this general acquiescence and consent for an argument of weight, as being the judgment of Time, let me tell him that the reasoning on which he relies is

[19]*Bacon* means that the mechanical arts progress from crude to better, but then are allowed to languish and are not driven to their full capabilities.
[20]*Roman* senators of inferior rank.
[21]*Making* over to a special owner.

most fallacious and weak. For, first, we are far from knowing all that in the matter of sciences and arts has in various ages and places been brought to light and published; much less, all that has been by private persons secretly attempted and stirred; so neither the births nor the miscarriages of Time are entered in our records. Nor, secondly, is the consent itself and the time it has continued a consideration of much worth. For however various are the forms of civil polities, there is but one form of polity in the sciences; and that always has been and always will be popular. Now the doctrines which find most favour with the populace are those which are either contentious and pugnacious, or specious and empty; such, I say, as either entangle[22] assent or tickle it. And therefore no doubt the greatest wits in each successive age have been forced out of their own course; men of capacity and intellect above the vulgar having been fain,[23] for reputation's sake, to bow to the judgment of the time and the multitude; and thus if any contemplations of a higher order took light anywhere, they were presently blown out by the winds of vulgar opinions. So that Time is like a river, which has brought down to us things light and puffed up, while those which are weighty and solid have sunk. Nay, those very authors who have usurped a kind of dictatorship in the sciences and taken upon them to lay down the law with such confidence, yet when from time to time they come to themselves again, they fall to complaints of the subtlety of nature, the hiding-places of truth, the obscurity of things, the entanglement of causes, the weakness of the human mind; wherein nevertheless they show themselves never the more modest, seeing that they will rather lay the blame upon the common condition of men and nature than upon themselves. And then whatever any art fails to attain, they ever set it down upon the authority of that art itself as impossible of attainment; and how can art be found guilty when it is judge in its own cause? So it is but a device for exempting ignorance from ignominy. Now for those things which are delivered and received, this is their

[22]*Catch*, ensnare.
[23]*Willing*, inclined, happy.

condition: barren of works, full of questions; in point of enlargement slow and languid; carrying a show of perfection in the whole, but in the parts ill filled up; in selection popular, and unsatisfactory even to those who propound them; and therefore fenced round and set forth with sundry artifices. And if there be any who have determined to make trial for themselves, and put their own strength to the work of advancing the boundaries of the sciences, yet have they not ventured to cast themselves completely loose from received opinions or to seek their knowledge at the fountain;[24] but they think they have done some great thing if they do but add and introduce into the existing sum of science something of their own; prudently considering with themselves that by making the addition they can assert their liberty, while they retain the credit of modesty by assenting to the rest. But these mediocrities and middle ways so much praised, in deferring to opinions and customs, turn to the great detriment of the sciences. For it is hardly possible at once to admire an author and to go beyond him; knowledge being as water, which will not rise above the level from which it fell. Men of this kind, therefore, amend some things, but advance little; and improve the condition of knowledge, but do not extend its range. Some, indeed, there have been who have gone more boldly to work, and taking it all for an open matter and giving their genius full play, have made a passage for themselves and their own opinions by pulling down and demolishing former ones; and yet all their stir has but little advanced the matter; since their aim has been not to extend philosophy and the arts in substance and value, but only to change doctrines and transfer the kingdom of opinions to themselves; whereby little has indeed been gained, for though the error be the opposite of the other, the causes of erring are the same in both. And if there have been any who, not binding themselves either to other men's opinions or to their own, but loving liberty, have desired to engage others along with themselves in search, these, though honest in intention, have been weak in endeavour. For they have been

[24]*The* source.

content to follow probable[25] reasons, and are carried round in a whirl of arguments, and in the promiscuous liberty of search have relaxed the severity of inquiry. There is none who has dwelt upon experience and the facts of nature as long as is necessary. Some there are indeed who have committed themselves to the waves of experience, and almost turned mechanics;[26] yet these again have in their very experiments pursued a kind of wandering inquiry, without any regular system of operations. And besides they have mostly proposed to themselves certain petty tasks, taking it for a great matter to work out some single discovery;—a course of proceeding at once poor in aim and unskilful in design. For no man can rightly and successfully investigate the nature of anything in the thing itself; let him vary his experiments as laboriously as he will, he never comes to a resting-place, but still finds something to seek beyond.[27] And there is another thing to be remembered; namely, that all industry in experimenting has begun with proposing to itself certain definite works to be accomplished, and has pursued them with premature and unseasonable eagerness; it has sought, I say, experiments of Fruit, not experiments of Light; not imitating the divine procedure, which in its first day's work created light only and assigned to it one entire day; on which day it produced no material work, but proceeded to that on the days following. As for those who have given the first place to Logic, supposing that the surest helps to the sciences were to be found in that, they have indeed most truly and excellently perceived that the human intellect left to its own course is not to be trusted; but then the remedy is altogether too weak for the disease; nor is it without evil in itself. For the Logic which is received, though it be very properly applied to civil business and to those arts which rest in discourse and opinion, is not nearly subtle enough to deal with nature; and in offering at[28] what it cannot master,

[25]*Only* apparently true, only likely.
[26]*Practitioners* or students of the practical arts and sciences.
[27]*Even* after varied experimentation there is always more to do.
[28]*Aiming* at.

has done more to establish and perpetuate error than to open the way to truth.

Upon the whole therefore, it seems that men have not been happy hitherto either in the trust which they have placed in others or in their own industry with regard to the sciences; especially as neither the demonstrations nor the experiments as yet known are much to be relied upon. But the universe to the eye of the human understanding is framed like a labyrinth; presenting as it does on every side so many ambiguities of way, such deceitful resemblances of objects and signs, natures so irregular in their lines, and so knotted and entangled. And then the way is still to be made by the uncertain light of the sense, sometimes shining out, sometimes clouded over, through the woods of experience and particulars; while those who offer themselves for guides are (as was said) themselves also puzzled, and increase the number of errors and wanderers. In circumstances so difficult neither the natural force of man's judgment nor even any accidental felicity[29] offers any chance of success. No excellence of wit, no repetition of chance experiments, can overcome such difficulties as these. Our steps must be guided by a clue, and the whole way from the very first perception of the senses must be laid out upon a sure plan. Not that I would be understood to mean that nothing whatever has been done in so many ages by so great labours. We have no reason to be ashamed of the discoveries which have been made, and no doubt the ancients proved themselves in everything that turns on wit and abstract meditation, wonderful men. But as in former ages when men sailed only by observation of the stars, they could indeed coast along the shores of the old continent or cross a few small and mediterranean[30] seas; but before the ocean could be traversed and the new world discovered, the use of the mariner's needle, as a more faithful and certain guide, had to be found out; in like manner the discoveries which have been hitherto made in the arts and sciences are such as might be made by practice, meditation, observation, argumentation,—

[29]*Accidental* good result.
[30]*Inland.*

for they lay near to the senses, and immediately beneath common notions; but before we can reach the remoter and more hidden parts of nature, it is necessary that a more perfect use and application of the human mind and intellect be introduced.

For my own part at least, in obedience to the everlasting love of truth, I have committed myself to the uncertainties and difficulties and solitudes of the ways, and relying on the divine assistance have upheld my mind both against the shocks and embattled ranks of opinion, and against my own private and inward hesitations and scruples, and against the fogs and clouds of nature, and the phantoms flitting about on every side; in the hope of providing at last for the present and future generations guidance more faithful and secure. Wherein if I have made any progress, the way has been opened to me by no other means than the true and legitimate humiliation of the human spirit. For all those who before me have applied themselves to the invention of arts have but cast a glance or two upon facts and examples and experience, and straightway proceeded, as if invention were nothing more than an exercise of thought, to invoke their own spirits to give them oracles. I, on the contrary, dwelling purely and constantly among the facts of nature, withdraw my intellect from them no further than may suffice to let the images and rays of natural objects meet in a point, as they do in the sense of vision; whence it follows that the strength and excellency of the wit has but little to do in the matter. And the same humility which I use in inventing I employ likewise in teaching. For I do not endeavour either by triumphs of confutation, or pleadings of antiquity, or assumption of authority, or even by the veil of obscurity, to invest these inventions of mine with any majesty; which might easily be done by one who sought to give lustre to his own name rather than light to other men's minds. I have not sought (I say) nor do I seek either to force or ensnare men's judgments, but I lead them to things themselves and the concordances[31] of things, that they may see for themselves what they have, what

---

[31] *Agreements*, harmonies, connections.

they can dispute, what they can add and contribute to the common stock. And for myself, if in anything I have been either too credulous or too little awake and attentive, or if I have fallen off by the way and left the inquiry incomplete, nevertheless I so present these things naked and open, that my errors can be marked and set aside before the mass of knowledge be further infected by them; and it will be easy also for others to continue and carry on my labours. And by these means I suppose that I have established for ever a true and lawful marriage between the empirical and the rational faculty, the unkind and ill-starred divorce and separation of which has thrown into confusion all the affairs of the human family.

Wherefore, seeing that these things do not depend upon myself, at the outset of the work I most humbly and fervently pray to God the Father, God the Son, and God the Holy Ghost, that remembering the sorrows of mankind and the pilgrimage of this our life wherein we wear out days few and evil, they will vouchsafe through my hands to endow the human family with new mercies. This likewise I humbly pray, that things human may not interfere with things divine, and that from the opening of the ways of sense[32] and the increase of natural light there may arise in our minds no incredulity or darkness with regard to the divine mysteries; but rather that the understanding being thereby purified and purged of fancies and vanity, and yet not the less subject and entirely submissive to the divine oracles, may give to faith that which is faith's. Lastly, that knowledge being now discharged of that venom which the serpent infused into it, and which makes the mind of man to swell, we may not be wise above measure and sobriety, but cultivate truth in charity.

And now having said my prayers I turn to men; to whom I have certain salutary admonitions to offer and certain fair requests to make. My first admonition (which was also my prayer) is that men confine the sense within the limits of duty in respect of things divine: for the sense is like the sun, which reveals the face of earth, but seals and shuts up the face of

[32]***The*** renewed investigation of empirical evidence.

heaven. My next, that in flying from this evil they fall not into the opposite error, which they will surely do if they think that the inquisition of nature is in any part interdicted or forbidden. For it was not that pure and uncorrupted natural knowledge whereby Adam gave names to the creatures according to their propriety,[33] which gave occasion to the fall. It was the ambitious and proud desire of moral knowledge to judge of good and evil, to the end that man may revolt from God and give laws to himself, which was the form and manner of the temptation. Whereas of the sciences which regard nature, the divine philosopher[34] declares that "it is the glory of God to conceal a thing, but it is the glory of the King to find a thing out."[35] Even as though the divine nature took pleasure in the innocent and kindly sport of children playing at hide and seek, and vouchsafed of his kindness and goodness to admit the human spirit for his playfellow at that game. Lastly, I would address one general admonition to all; that they consider what are the true ends of knowledge, and that they seek it not either for pleasure of the mind, or for contention, or for superiority to others, or for profit, or fame, or power, or any of these inferior things; but for the benefit and use of life; and that they perfect and govern it in charity. For it was from lust of power that the angels fell, from lust of knowledge that man fell; but of charity there can be no excess, neither did angel or man ever come in danger by it.

The requests I have to make are these. Of myself I say nothing; but in behalf of the business which is in hand I entreat men to believe that it is not an opinion to be held, but a work to be done; and to be well assured that I am labouring to lay the foundation, not of any sect or doctrine, but of human utility and power. Next, I ask them to deal fairly by their own interests, and laying aside all emulations[36] and prejudices in favor of this or that opinion, to join in consultation for the

---

[33]*Particular* nature or character.
[34]*Solomon.*
[35]*Proverbs* 25: 2.
[36]*Ambitious* rivalries for power.

common good; and being now freed and guarded by the securities and helps which I offer from the errors and impediments of the way, to come forward themselves and take part in that which remains to be done. Moreover, to be of good hope, nor to imagine that this Instauration of mine is a thing infinite and beyond the power of man, when it is in fact the true end and termination of infinite error; and seeing also that it is by no means forgetful of the conditions of mortality and humanity, (for it does not suppose that the work can be altogether completed within one generation, but provides for its being taken up by another); and finally that it seeks for the sciences not arrogantly in the little cells of human wit, but with reverence in the greater world. But it is the empty things that are vast: things solid are most contracted and lie in little room.[37] And now I have only one favour more to ask (else injustice to me may perhaps imperil the business itself)—that men will consider well how far, upon that which I must needs assert (if I am to be consistent with myself), they are entitled to judge and decide upon these doctrines of mine; inasmuch as all that premature human reasoning which anticipates inquiry, and is abstracted from the facts rashly and sooner than is fit, is by me rejected (so far as the inquisition of nature is concerned), as a thing uncertain, confused, and ill built up; and I cannot be fairly asked to abide by the decision of a tribunal which is itself on its trial.

[37]*The* task of the Instauration is possible.

The Plan of the Work

The Work Is In Six Parts:—

1. *The Divisions of the Sciences.*
2. *The New Organon*[38]*; or Directions concerning the Interpretation of Nature.*
3. *The Phenomena of the Universe; or a Natural and Experimental History for the foundation of Philosophy.*
4. *The Ladder of the Intellect.*
5. *The Forerunners; or Anticipations of the New Philosophy.*
6. *The New Philosophy; or Active Science.*

The Arguments of the several Parts.

It being part of my design to set everything forth, as far as may be, plainly and perspicuously (for nakedness of the mind is still, as nakedness of the body once was, the companion of innocence and simplicity), let me first explain the order and plan of the work. I distribute it into six parts.

The first part[39] exhibits a summary or general description of the knowledge which the human race at present possesses. For I thought it good to make some pause upon that which is received; that thereby the old may be more easily made perfect and the new more easily approached. And I hold the improvement of that which we have to be as much an

---

[38]*An* instrument of thought or knowledge; also the traditional title of Aristotle's logical treatises.

[39]*Part* one is elsewhere declared to be wanting although "some account [of the divisions of the sciences] will be found in the 'Second Book of the Proficience and Advancement of Learning' "; below, p. 33.

New Atlantis and The Great Instauration, Second Edition.
Edited by Jerry Weinberger.
© 2017 John Wiley & Sons, Inc. Published 2017 by John Wiley & Sons, Inc.

object as the acquisition of more. Besides which it will make me the better listened to; for "He that is ignorant (says the proverb) receives not the words of knowledge, unless thou first tell him that which is in his own heart." We will therefore make a coasting voyage along the shores of the arts and sciences received, not without importing into them some useful things by the way.

In laying out the divisions of the sciences however, I take into account not only things already invented and known, but likewise things omitted which ought to be there. For there are found in the intellectual as in the terrestial globe waste regions as well as cultivated ones. It is no wonder therefore if I am sometimes obliged to depart from the ordinary divisions. For in adding to the total you necessarily alter the parts and sections; and the received divisions of the sciences are fitted only to the received sum of them as it stands now.

With regard to those things which I shall mark as omitted, I intend not merely to set down a simple title or a concise argument of that which is wanted. For as often as I have occasion to report anything as deficient, the nature of which is at all obscure, so that men may not perhaps easily understand what I mean or what the work is which I have in my head, I shall always (provided it be a matter of any worth) take care to subjoin either directions for the execution of such work, or else a portion of the work itself executed by myself as a sample of the whole: thus giving assistance in every case either by work or by counsel. For if it were for the sake of my own reputation only and other men's interests were not concerned in it, I would not have any man think that in such cases merely some light and vague notion has crossed my mind, and that the things which I desire and offer at are no better than wishes; when they are in fact things which men may certainly command if they will, and of which I have formed in my own mind a clear and detailed conception. For I do not propose merely to survey these regions in my mind, like an augur taking auspices, but to enter them like a general who means to take possession.—So much for the first part of the work.

Having thus coasted past the ancient arts, the next point is to equip the intellect for passing beyond. To the second part[40] therefore belongs the doctrine concerning the better and more perfect use of human reason in the inquisition of things, and the true helps of the understanding: that thereby (as far as the condition of mortality and humanity allows) the intellect may be raised and exalted, and made capable of overcoming the difficulties and obscurities of nature. The art which I introduce with this view (which I call *Interpretation of Nature*) is a kind of logic; though the difference between it and the ordinary logic is great; indeed immense. For the ordinary logic professes to contrive and prepare helps and guards for the understanding, as mine does; and in this one point they agree. But mine differs from it in three points especially; viz. in the end aimed at; in the order of demonstration; and in the starting point of the inquiry.

For the end which this science of mine proposes is the invention not of arguments but of arts; not of things in accordance with principles, but of principles themselves; not of probable reasons, but of designations and directions for works. And as the intention is different, so accordingly is the effect; the effect of the one being to overcome an opponent in argument, of the other to command[41] nature in action.

In accordance with this end is also the nature and order of the demonstrations. For in the ordinary logic almost all the work is spent about the syllogism.[42] Of induction[43] the logicians seem hardly to have taken any serious thought, but they pass it by with a slight notice, and hasten on to the formulæ of disputation. I on the contrary reject demonstration by syllogism, as acting too confusedly, and letting nature slip out of its hands. For although no one can doubt that things which agree

[40]*The* second part is provided in the *Novum organum*, but only as "a summary digested into aphorisms"; ibid.

[41]*Rees* translates this as "to bend nature to works." See Rees, 29.

[42]*The* scheme of deductive logic composed of a major premise (every animal is mortal), a minor premise (every man is an animal), and a conclusion (every man is mortal).

[43]*Reasoning* from particulars to general principles.

in a middle term[44] agree with one another (which is a proposition of mathematical certainty), yet it leaves an opening for deception; which is this. The syllogism consists of propositions; propositions of words; and words are the tokens and signs of notions. Now if the very notions of the mind (which are as the soul of words and the basis of the whole structure) be improperly and overhastily abstracted from facts, vague, not sufficiently definite, faulty in short in many ways, the whole edifice tumbles. I therefore reject the syllogism; and that not only as regards principles (for to principles the logicians themselves do not apply it) but also as regards middle propositions;[45] which, though obtainable no doubt by the syllogism, are, when so obtained, barren of works, remote from practice, and altogether unavailable for the active department of the sciences. Although therefore I leave to the syllogism and these famous and boasted modes of demonstration their jurisdiction over popular arts and such as are matter of opinion (in which department I leave all as it is), yet in dealing with the nature of things I use induction throughout, and that in the minor propositions[46] as well as the major. For I consider induction to be that form of demonstration which upholds the sense, and closes[47] with nature, and comes to the very brink of operation, if it does not actually deal with it.

Hence it follows that the order of demonstration is likewise inverted. For hitherto the proceeding has been to fly at once from the sense[48] and particulars up to the most general propositions, as certain fixed poles for the argument to turn upon, and from these to derive the rest by middle terms: a short way, no doubt, but precipitate; and one which will never lead

---

[44]*The* term common to the two premises of a syllogism, *i.e.*, "animal" in the example in n. 42, above.

[45]*Propositions* about the middle term of a syllogism. In a syllogism, the middle term disappears from the conclusion.

[46]*Premises.* Bacon means here that he will apply inductive reasoning to the propositions of major and minor premises.

[47]*Comes* into contact with

[48]*Empirical* evidence.

to nature, though it offers an easy and ready way to disputation. Now my plan is to proceed regularly and gradually from one axiom to another, so that the most general are not reached till the last: but then when you do come to them you find them to be not empty notions, but well defined, and such as nature would really recognise as her first principles, and such as lie at the heart and marrow of things.

But the greatest change I introduce is in the form itself of induction and the judgment made thereby. For the induction of which the logicians speak, which proceeds by simple enumeration, is a puerile thing; concludes at hazard; is always liable to be upset by a contradictory instance; takes into account only what is known and ordinary; and leads to no result.

Now what the sciences stand in need of is a form of induction which shall analyse experience and take it to pieces, and by a due process of exclusion and rejection lead to an inevitable conclusion. And if that ordinary mode of judgment practised by the logicians was so laborious, and found exercise for such great wits, how much more labour must we be prepared to bestow upon this other, which is extracted not merely out of the depths of the mind, but out of the very bowels of nature.

Nor is this all. For I also sink the foundations of the sciences deeper and firmer; and I begin the inquiry nearer the source than men have done heretofore; submitting to examination those things which the common logic takes on trust. For first, the logicians borrow the principles of each science from the science itself; secondly, they hold in reverence the first notions of the mind; and lastly, they receive as conclusive the immediate informations of the sense, when well disposed. Now upon the first point, I hold that true logic ought to enter the several provinces of science armed with a higher authority than belongs to the principles of those sciences themselves, and ought to call those putative principles to account until they are fully established. Then with regard to the first notions of the intellect; there is not one of the impressions taken by the intellect when left to go its own way, but I hold it for suspected, and no way established, until it has submitted to a new trial and a fresh judgment has been thereupon pronounced. And lastly, the information of the sense itself I sift and examine in many ways. For certain it is that the senses deceive; but then at the same time they supply the means of discovering their own errors; only the errors are here, the means of discovery are to seek.

The sense fails in two ways. Sometimes it gives no information, sometimes it gives false information. For first, there are very many things which escape the sense, even when best disposed and no way obstructed; by reason either of the subtlety of the whole body, or the minuteness of the parts, or distance of place, or slowness or else swiftness of motion, or familiarity of the object, or other causes. And again when the sense does apprehend a thing its apprehension is not much to be relied upon. For the testimony and information of the sense has reference always to man, not to the universe; and it is a great error to assert that the sense is the measure of things.

To meet these difficulties, I have sought on all sides diligently and faithfully to provide helps for the sense—substitutes to supply its failures, rectifications to correct its errors; and this I endeavour to accomplish not so much by instruments as by experiments. For the subtlety of experiments is far greater than that of the sense itself, even when assisted by exquisite instruments; such experiments, I mean, as are skilfully and artificially devised for the express purpose of determining the point in question. To the immediate and proper perception of the sense therefore I do not give much weight; but I contrive that the office[49] of the sense shall be only to judge of the experiment, and that the experiment itself shall judge of the thing. And thus I conceive that I perform the office of a true priest of the sense (from which all knowledge in nature must be sought, unless men mean to go mad) and a not unskilful interpreter of its oracles; and that while others only profess to uphold and cultivate the sense, I do so in fact. Such then are the provisions I make for finding the genuine light of nature and kindling and bringing it to bear. And they would be sufficient of themselves, if the human intellect were even, and like a fair sheet of paper with no writing on it. But since the minds of men are strangely

[49] *Function.*

possessed and beset, so that there is no true and even surface left to reflect the genuine rays of things, it is necessary to seek a remedy for this also.

Now the idols, or phantoms, by which the mind is occupied are either adventitious or innate. The adventitious come into the mind from without; namely, either from the doctrines and sects of philosophers, or from perverse rules of demonstration. But the innate are inherent in the very nature of the intellect, which is far more prone to error than the sense is. For let men please themselves as they will in admiring and almost adoring the human mind, this is certain: that as an uneven mirror distorts the rays of objects according to its own figure and section, so the mind, when it receives impressions of objects through the sense, cannot be trusted to report them truly, but in forming its notions mixes up its own nature with the nature of things.

And as the first two kinds of idols are hard to eradicate, so idols of this last kind cannot be eradicated at all. All that can be done is to point them out, so that this insidious action of the mind may be marked and reproved (else as fast as old errors are destroyed new ones will spring up out of the ill complexion[50] of the mind itself, and so we shall have but a change of errors, and not a clearance); and to lay it down once for all as a fixed and established maxim, that the intellect is not qualified to judge except by means of induction, and induction in its legitimate form. This doctrine then of the expurgation of the intellect to qualify it for dealing with truth, is comprised in three refutations: the refutation of the Philosophies; the refutation of the Demonstrations; and the refutation of the Natural Human Reason.[51] The explanation of which things, and of the

---

[50]*Nature*, disposition.
[51]*Bacon's* considerations of these "refutations" are scattered throughout his writings. See Anderson, *The Philosophy of Francis Bacon*, pp. 40–47. For the "refutation of the Philosophies," see Farrington, *The Philosophy of Francis Bacon*, which contains translations of *The Masculine Birth of Time*, *Thoughts and Conclusions*, and *The Refutation of Philosophies*.

true relation between the nature of things and the nature of the mind, is as the strewing and decoration of the bridal chamber of the Mind and the Universe, the Divine Goodness assisting; out of which marriage let us hope (and be this the prayer of the bridal song) there may spring helps to man, and a line and race of inventions that may in some degree subdue and overcome the necessities and miseries of humanity. This is the second part of the work.

But I design not only to indicate and mark out the ways, but also to enter them. And therefore the third part[52] of the work embraces the Phenomena of the Universe; that is to say, experience of every kind, and such a natural history as may serve for a foundation to build philosophy upon. For a good method of demonstration or form of interpreting nature may keep the mind from going astray or stumbling, but it is not any excellence of method that can supply it with the material of knowledge. Those however who aspire not to guess and divine, but to discover and know; who propose not to devise mimic[53] and fabulous worlds of their own, but to examine and dissect the nature of this very world itself; must go to facts themselves for everything. Nor can the place of this labour and search and worldwide perambulation be supplied by any genius or meditation or argumentation; no, not if all men's wits could meet in one. This therefore we must have, or the business must be for ever abandoned. But up to this day such has been the condition of men in this matter, that it is no wonder if nature will not give herself into their hands.

For first, the information of the sense itself, sometimes failing, sometimes false; observation; careless, irregular, and led by chance; tradition, vain and fed on rumour; practice, slavishly bent upon its work; experiment, blind, stupid, vague, and prematurely broken off; lastly, natural history trivial and poor;—all these have contributed to supply the understanding with very bad materials for philosophy and the sciences.

[52]*See* Rees, xix–xxii.
[53]*Imitation.*

Then an attempt is made to mend the matter by a preposterous subtlety and winnowing of argument. But this comes too
late, the case being already past remedy; and is far from setting
the business right or sifting away the errors. The only hope
therefore of any greater increase or progress lies in a
reconstruction of the sciences.

Of this reconstruction the foundation must be laid in natural
history, and that of a new kind and gathered on a new
principle. For it is in vain that you polish the mirror if there
are no images to be reflected; and it is as necessary that the
intellect should be supplied with fit matter to work upon, as
with safeguards to guide its working. But my history differs
from that in use (as my logic does) in many things,—in end and
office, in mass and composition, in subtlety, in selection also
and setting forth, with a view to the operations which are to
follow.

For first, the object of the natural history which I propose is
not so much to delight with variety of matter or to help with
present use of experiments, as to give light to the discovery of
causes and supply a suckling philosophy with its first food. For
though it be true that I am principally in pursuit of works and
the active department of the sciences, yet I wait for harvest-
time, and do not attempt to mow the moss[54] or to reap the
green corn. For I well know that axioms once rightly discovered will carry whole troops of works along with them, and
produce them, not here and there one, but in clusters. And that
unseasonable and puerile hurry to snatch by way of earnest at
the first works which come within reach, I utterly condemn and
reject, as an Atalanta's apple[55] that hinders the race. Such then
is the office of this natural history of mine.

Next, with regard to the mass and composition of it: I mean
it to be a history not only of nature free and at large (when she
is left to her own course and does her work her own way),—

---

[54]*Cotton* grass.
[55]*In* Greek mythology, Hippomenes won the swift Atalanta's hand by
defeating her in a footrace. At the advice of Aphrodite, Hippomenes
slowed Atlanta by dropping the three apples of the Hesperides.

such as that of the heavenly bodies, meteors, earth and sea, minerals, plants, animals,—but much more of nature under constraint and vexed; that is to say, when by art and the hand of man she is forced out of her natural state, and squeezed and moulded. Therefore I set down at length all experiments of the mechanical arts, of the operative part of the liberal arts, of the many crafts which have not yet grown into arts properly so called, so far as I have been able to examine them and as they conduce to the end in view. Nay (to say the plain truth) I do in fact (low and vulgar as men may think it) count more upon this part both for helps and safeguards than upon the other; seeing that the nature of things betrays itself more readily under the vexations of art than in its natural freedom.

Nor do I confine the history to Bodies; but I have thought it my duty besides to make a separate history of such Virtues[56] as may be considered cardinal in nature. I mean those original passions[57] or desires[58] of matter which constitute the primary elements of nature; such as Dense and Rare, Hot and Cold, Solid and Fluid, Heavy and Light, and several others.[59]

Then again, to speak of subtlety: I seek out and get together a kind of experiments much subtler and simpler than those which occur accidentally.[60] For I drag into light many things which no one who was not proceeding by a regular and certain way to the discovery of causes would have thought of inquiring after; being indeed in themselves of no great use; which shows that they were not sought for on their own account; but having just the same relation to things and works which the letters of the alphabet have to speech and words—which, though in themselves useless, are the elements of which all discourse is made up.

Further, in the selection of the relations and experiments I conceive I have been a more cautious purveyor than those who have hitherto dealt with natural history. For I admit nothing

---

[56]*Qualities*, especially qualities that are beneficial to the human body.
[57]*The* way a thing may be affected by an external agency.
[58]*Tendencies*.
[59]*See* Anderson, *The Philosophy of Francis Bacon*, pp. 34–35.
[60]*Casually*.

but on the faith of eyes, or at least of careful and severe examination; so that nothing is exaggerated for wonder's sake, but what I state is sound and without mixture of fables or vanity. All received or current falsehoods also (which by strange negligence have been allowed for many ages to prevail and become established) I proscribe and brand by name; that the sciences may be no more troubled with them. For it has been well observed that the fables and superstitions and follies which nurses instil into children do serious injury to their minds; and the same consideration makes me anxious, having the management of the childhood as it were of philosophy in its course of natural history, not to let it accustom itself in the beginning to any vanity. Moreover, whenever I come to a new experiment of any subtlety (though it be in my own opinion certain and approved), I nevertheless subjoin a clear account of the manner in which I made it; that men knowing exactly how each point was made out, may see whether there be any error connected with it, and may arouse themselves to devise proofs more trustworthy and exquisite, if such can be found; and finally, I interpose everywhere admonitions and scruples and cautions, with a religious care to eject, repress, and as it were exorcise every kind of phantasm.

Lastly, knowing how much the sight of man's mind is distracted by experience and history, and how hard it is at the first (especially for minds either tender or preoccupied) to become familiar with nature, I not unfrequently subjoin observations of my own, being as the first offers, inclinations, and as it were glances of history towards philosophy; both by way of an assurance to men that they will not be kept for ever tossing on the waves of experience, and also that when the time comes for the intellect to begin its work, it may find everything the more ready. By such a natural history then as I have described, I conceive that a safe and convenient approach may be made to nature, and matter supplied of good quality and well prepared for the understanding to work upon.

And now that we have surrounded the intellect with faithful helps and guards, and got together with most careful selection a regular army of divine works, it may seem that we have no more to do but to proceed to philosophy itself. And yet in a matter so difficult and doubtful there are still some things which it seems necessary to premise,[61] partly for convenience of explanation, partly for present use.

Of these the first is to set forth examples of inquiry and invention according to my method, exhibited by anticipation in some particular subjects; choosing such subjects as are at once the most noble in themselves among those under inquiry, and most different one from another; that there may be an example in every kind. I do not speak of those examples which are joined to the several precepts and rules by way of illustration (for of these I have given plenty in the second part of the work); but I mean actual types and models, by which the entire process of the mind and the whole fabric and order of invention from the beginning to the end, in certain subjects, and those various and remarkable, should be set as it were before the eyes. For I remember that in the mathematics it is easy to follow the demonstration when you have a machine beside you; whereas without that help all appears involved and more subtle than it really is. To examples of this kind,—being in fact nothing more than an application of the second part in detail and at large,— the fourth part[62] of the work is devoted.

The fifth part[63] is for temporary use only, pending the completion of the rest; like interest payable from time to time until the principal be forthcoming. For I do not make so blindly for the end of my journey, as to neglect anything useful that may turn up by the way. And therefore I include in this fifth part such things as I have myself discovered, proved, or added,—not however according to the true rules and methods of interpretation, but by the ordinary use of the understanding in inquiring and discovering. For besides that I hope my speculations may in virtue of my continual conversancy with nature have a value beyond the pretensions of my

---

[61]*To* set forth beforehand.
[62]*Bacon* completed only a very small fragment of the fourth part.
[63]*Bacon* completed only a very small fragment of the fifth part.

wit, they will serve in the meantime for wayside inns, in which the mind may rest and refresh itself on its journey to more certain conclusions. Nevertheless I wish it to be understood in the meantime that they are conclusions by which (as not being discovered and proved by the true form of interpretation) I do not at all mean to bind myself. Nor need any one be alarmed at such suspension of judgment, in one who maintains not simply that nothing can be known, but only that nothing can be known except in a certain course and way; and yet establishes provisionally certain degrees of assurance, for use and relief until the mind shall arrive at a knowledge of causes in which it can rest. For even those schools of philosophy which held the absolute impossibility of knowing anything were not inferior to those which took upon them to pronounce. But then they did not provide helps for the sense and understanding, as I have done, but simply took away all their authority: which is quite a different thing—almost the reverse.

The sixth part[64] of my work (to which the rest is subservient and ministrant)[65] discloses and sets forth that philosophy which by the legitimate, chaste, and severe course of inquiry which I have explained and provided is at length developed and established. The completion however of this last part is a thing both above my strength and beyond my hopes. I have made a beginning of the work—a beginning, as I hope, not unimportant:—the fortune of the human race will give the issue;—such an issue, it may be, as in the present condition of things and men's minds cannot easily be conceived or imagined. For the matter in hand is no mere felicity[66] of speculation, but the real business and fortunes of the human race, and all power of operation. For man is but the servant and interpreter of nature: what he does and what he knows is only what he has observed of nature's order in fact or in thought; beyond this he knows nothing and can do nothing. For the chain of causes

---

[64]*As* a formally identified part of the plan, the sixth part does not exist in Bacon's writing. See pp. xviii–xx, above.
[65]*A* minister to.
[66]*Happy* expression.

cannot by any force be loosed or broken, nor can nature be commanded[67] except by being obeyed. And so those twin objects, human Knowledge and human Power, do really meet in one; and it is from ignorance of causes that operation fails.

And all depends on keeping the eye steadily fixed upon the facts of nature and so receiving their images simply as they are. For God forbid that we should give out a dream of our own imagination for a pattern of the world; rather may he graciously grant to us to write an apocalypse[68] or true vision of the footsteps of the Creator imprinted on his creatures.

Therefore do thou, O Father, who gavest the visible light as the first fruits of creation, and didst breathe into the face of man the intellectual light as the crown and consummation thereof, guard and protect this work, which coming from thy goodness returneth to thy glory. Thou when thou turnedst to look upon the works which thy hands had made, sawest that all was very good, and didst rest from thy labours. But man, when he turned to look upon the work which his hands had made, saw that all was vanity and vexation of spirit, and could find no rest therein. Wherefore if we labour in thy works with the sweat of our brows thou wilt make us partakers of thy vision and thy sabbath. Humbly we pray that this mind may be steadfast in us, and that through these our hands, and the hands of others to whom thou shalt give the same spirit, thou wilt vouchsafe to endow the human family with new mercies.

[67] *The* Latin is *vincitur*, which could be translated "conquered" or "overcome." See Rees, 44–45.
[68] *Prophetic* revelation.

The

FIRST PART OF THE INSTAURATION,

Which comprises the

DIVISIONS OF THE SCIENCES,

IS WANTING.

But some account of them will be found in the
Second
Book of the
"Proficience and Advancement of Learning,
Divine and Human."

Next comes

The

SECOND PART OF THE INSTAURATION,

WHICH EXHIBITS

THE ART ITSELF OF INTERPRETING NATURE,

AND OF THE TRUER EXERCISE OF THE INTELLECT;

Not however in the form of a regular Treatise, but only a
Summary
digested into Aphorisms.[1]

---

[1]*The* preceding titles (Prooemium, Epistle Dedicatory, Preface, and
Plan of the Work) constitute the preliminaries of the project Bacon
entitled The Great Instauration and published in the 1620 *Novum
organum*. This page, coming just before *Novum organum*, makes it
clear that the first part of the plan, the division of the sciences, is only
partly complete as the 1605 "Proficience and Advancement of Learning" (although that part was greatly expanded and so completed in the
later 1623 *De augmentis scientiarum*). Bacon here tells us that the
second part of the plan, the true art of interpreting nature (*Novum
organum*), is in the form of two books of aphorisms (terse expressions
of philosophic or scientific principles). In the first book of aphorisms,
the Idols of the Mind are outlined in aphorisms XXXVIII through
LXVIII.

THE IDOLS
OF THE MIND

XXXVIII

The idols and false notions which are now in possession of
the human understanding, and have taken deep root therein,
not only so beset men's minds that truth can hardly find
entrance, but even after entrance obtained, they will again
in the very instauration of the sciences meet and trouble us,
unless men being forewarned of the danger fortify themselves
as far as may be against their assaults.

XXXIX

There are four classes of Idols which beset men's minds. To
these for distinction's sake I have assigned names,—calling the
first class *Idols of the Tribe*; the second, *Idols of the Cave*; the
third, *Idols of the Market-Place*; the fourth, *Idols of the Theater*.

XL

The formation of ideas and axioms by true induction is no
doubt the proper remedy to be applied for the keeping off and
clearing away of idols. To point them out, however, is of great use;
for the doctrine of Idols is to the Interpretation of Nature what the
doctrine of the refutation of Sophisms is to common Logic.

New Atlantis and The Great Instauration, Second Edition.
Edited by Jerry Weinberger.
© 2017 John Wiley & Sons, Inc. Published 2017 by John Wiley & Sons, Inc.

XLI

The Idols of the Tribe have their foundation in human nature itself, and in the tribe or race of men. For it is a false assertion that the sense of man is the measure of things.[2] On the contrary, all perceptions as well of the sense as of the mind are according to the measure of the individual and not according to the measure of the universe. And the human understanding is like a false mirror, which, receiving rays irregularly, distorts and discolors the nature of things by mingling its own nature with it.

XLII

The Idols of the Cave are the idols of the individual man. For every one (besides the errors common to human nature in general) has a cave or den of his own, which refracts and discolors the light of nature; owing either to his own proper and peculiar nature; or to his education and conversation with others; or to the reading of books, and the authority of those whom he esteems and admires; or to the differences of impressions, accordingly as they take place in a mind preoccupied and predisposed or in a mind indifferent and settled; or the like. So that the spirit of man (according as it is meted out to different individuals) is in fact a thing variable and full of perturbation, and governed as it were by chance. Whence it was well observed by Heraclitus that men look for sciences in their own lesser worlds, and not in the greater or common world.[3]

---

[2]*Bacon* could be thinking of the saying attributed to the Pre-Socratic philosopher Protagoras (c. 490–420 B.C.E.) "man is the measure of all things." See Diogenes Laertius, *Lives of Eminent Philosophers*, IX: 50–56.

[3]*Heraclitus* (c. 535–475 B.C.E.) was a Pre-Socratic Philosopher. See Sextus Empiricus, *Against the Logicians* I: 134; and Rees, 509.

XLIII

There are also Idols formed by the intercourse and association of men with each other, which I call Idols of the Marketplace, on account of the commerce and consort of men there. For it is by discourse that men associate; and words are imposed according to the apprehension of the vulgar. And therefore the ill and unfit choice of words wonderfully obstructs the understanding. Nor do the definitions or explanations wherewith in some things learned men are wont to guard and defend themselves, by any means set the matter right. But words plainly force and overrule the understanding, and throw all into confusion, and lead men away into numberless empty controversies and idle fancies.

XLIV

Lastly, there are Idols which have immigrated into men's minds from the various dogmas of philosophies, and also from wrong laws of demonstration. These I call Idols of the Theater; because in my judgment all the received systems are but so many stage-plays, representing worlds of their own creation after an unreal and scenic fashion. Nor is it only of the systems now in vogue, or only of the ancient sects and philosophies, that I speak; for many more plays of the same kind may yet be composed and in like artificial manner set forth; seeing that errors the most widely different have nevertheless causes for the most part alike. Neither again do I mean this only of entire systems, but also of many principles and axioms in science, which by tradition, credulity, and negligence have come to be received.

But of these several kinds of Idols I must speak more largely and exactly, that the understanding may be duly cautioned.

XLV

The human understanding is of its own nature prone to suppose the existence of more order and regularity in the world than it finds. And though there be many things in nature which are singular and unmatched, yet it devises for them parallels and conjugates and relatives which do not exist. Hence the fiction that all celestial bodies move in perfect circles; spirals and dragons being (except in name) utterly rejected.[4] Hence too the element of Fire with its orb is brought in, to make up the square with the other three which the sense perceives.[5] Hence also the ratio of density of the so-called elements is arbitrarily fixed at ten to one. And so on of other dreams. And these fancies affect not dogmas only, but simple notions also.

XLVI

The human understanding when it has once adopted an opinion (either as being the received opinion or as being agreeable to itself) draws all things else to support and agree with it. And though there be a greater number and weight of instances to be found on the other side, yet these it either neglects and despises, or else by some distinction sets aside and rejects; in order that by this great and pernicious pre-determination the authority of its former conclusions may

[4]*Bacon* is here arguing that irregular astronomical motions should be explained as spiral or irregular lines, not perfect circles. In BW, the following note is appended to the Latin text at this point: "It does not appear in what sense Bacon uses the word 'draco.' In its ordinary acceptation in old astronomy, it denoted the great circle which is approximately the projection on the sphere of the moon's orbit. The ascending node was called the caput draconis [dragon's head] and the descending the cauda draconis [dragon's tail]. The same terms were occasionally applied to the modes of the planetary orbits. It is not improbable that Bacon intended to complain of the rejection of spirals of double curvatures, or helices, which traced on the surface of the sphere might represent inequalities in latitude." BW, Vol. 1, 165. See also Rees, 563.

[5]*Bacon* is here objecting to the Aristotelian Coimbrans, professors at the Colegio das Artas in Coimbra, Portugal, who argued for a sphere of fire below the moon because the existence of the other three elements (earth, water, and air) necessitated it. See Rees, 510.

remain inviolate. And therefore it was a good answer that was made by one who when they showed him hanging in a temple a picture of those who had paid their vows as having escaped shipwreck, and would have him say whether he did not now acknowledge the power of the gods,—"Aye," asked he again, "but where are they painted that were drowned after their vows?"[6] And such is the way of all superstition, whether in astrology, dreams, omens, divine judgments, or the like; wherein men, having a delight in such vanities, mark the events where they are fulfilled, but where they fail, though this happen much oftener, neglect and pass them by. But with far more subtlety does this mischief insinuate itself into philosophy and the sciences; in which the first conclusion colours and brings into conformity with itself all that come after, though far sounder and better. Besides, independently of that delight and vanity which I have described, it is the peculiar and perpetual error of the human intellect to be more moved and excited by affirmatives than by negatives; whereas it ought properly to hold itself indifferently disposed toward both alike. Indeed in the establishment of any true axiom, the negative instance is the more forcible of the two.

XLVII

The human understanding is moved by those things most which strike and enter the mind simultaneously and suddenly, and so fill the imagination; and then it feigns and supposes all other things to be somehow, though it cannot see how, similar to those few things by which it is surrounded. But for that going to and fro to remote and heterogeneous instances, by which axioms are tried as in the fire, the intellect is altogether slow and unfit, unless it be forced thereto by severe laws and over-ruling authority.

[6]*Cicero De natura deorum*, 3: 37; See Diogenes Laertius, *Lives of Eminent Philosophers*, VI: 59; Rees, 511.

XLVIII

The human understanding is unquiet; it cannot stop or rest, and still presses onward, but in vain. Therefore it is that we cannot conceive of any end or limit to the world; but always as of necessity it occurs to us that there is something beyond. Neither again can it be conceived how eternity has flowed down to the present day; for that distinction which is commonly received of infinity in time past and in time to come can by no means hold; for it would thence follow that one infinity is greater than another, and that infinity is wasting away and tending to become finite. The like subtlety arises touching the infinite divisibility of lines, from the same inability of thought to stop. But this inability interferes more mischievously in the discovery of causes: for although the most general principles in nature ought to be held merely positive, as they are discovered, and cannot with truth be referred to a cause; nevertheless the human understanding being unable to rest still seeks something prior in the order of nature. And then it is that in struggling toward that which is further off it falls back upon that which is more nigh at hand; namely, on final causes: which have relation clearly to the nature of man rather than to the nature of the universe; and from this source have strangely defiled philosophy. But he is no less an unskilled and shallow philosopher who seeks causes of that which is most general, than he who in things subordinate and subaltern omits to do so.

XLIX

The human understanding is no dry light, but receives an infusion from the will and affections; whence proceed sciences which may be called "sciences as one would." For what a man had rather were true he more readily believes. Therefore he rejects difficult things from impatience of research; sober things, because they narrow hope; the deeper things of nature, from superstition; the light of experience, from arrogance and pride, lest his mind should seem to be occupied with things mean and transitory; things not commonly believed, out of deference to the opinion of the vulgar. Numberless in short are the ways, and sometimes imperceptible, in which the affections colour and infect the understanding.

L

But by far the greatest hindrance and aberration of the human understanding proceeds from the dulness, incompetency, and deceptions of the senses; in that things which strike the sense outweigh things which do not immediately strike it, though they be more important. Hence it is that speculation commonly ceases where sight ceases; insomuch that of things invisible there is little or no observation. Hence all the working of the spirits enclosed in tangible bodies lies hid and unobserved of men. So also all the more subtle changes of form in the parts of coarser substances (which they commonly call alteration, though it is in truth local motion through exceedingly small spaces) is in like manner unobserved. And yet unless these two things just mentioned be searched out and brought to light, nothing great can be achieved in nature, as far as the production of works is concerned. So again the essential nature of our common air, and of all bodies less dense than air (which are very many), is almost unknown. For the sense by itself is a thing infirm and erring; neither can instruments for enlarging or sharpening the senses do much; but all the truer kind of interpretation of nature is effected by instances and experiments fit and apposite; wherein the sense decides touching the experiment only, and the experiment touching the point in nature and the thing itself.

LI

The human understanding is of its own nature prone to abstractions and gives a substance and reality to things which are fleeting. But to resolve nature into abstractions is less to our purpose than to dissect her into parts; as did the school of Democritus,[7] which went further into nature than the rest.

---

[7] *Democritus* (? 460–357 B.C.E.) was a Pre-Socratic atomist philosopher. See Diogenes Laertius, *Lives of Eminent Philosophers*, IX: 34–49.

Matter rather than forms should be the object of our attention, its configurations and changes of configuration, and simple action, and law of action or motion; for forms are figments of the human mind, unless you will call those laws of action forms.

LII

Such then are the idols which I call *Idols of the Tribe*; and which take their rise either from the homogeneity of the substance of the human spirit, or from its preoccupation, or from its narrowness, or from its restless motion, or from an infusion of the affections, or from the incompetency of the senses, or from the mode of impression.

LIII

The *Idols of the Cave* take their rise in the peculiar constitution, mental or bodily, of each individual; and also in education, habit, and accident. Of this kind there is a great number and variety; but I will instance those the pointing out of which contains the most important caution, and which have most effect in disturbing the clearness of the understanding.

LIV

Men become attached to certain particular sciences and speculations, either because they fancy themselves the authors and inventors thereof, or because they have bestowed the greatest pains upon them and become most habituated to them. But men of this kind, if they betake themselves to philosophy and contemplations of a general character, distort and colour them in obedience to their former fancies; a thing especially to be noticed in Aristotle, who made his natural philosophy a mere bond-servant to his logic, thereby rendering it contentious and well nigh useless. The race of chemists again out of a few experiments of the furnace have built up a fantastic philosophy, framed with reference to a few things; and Gilbert[8] also, after he had employed himself most laboriously in the study and observation of the loadstone, proceeded at once to construct an entire system in accordance with his favorite subject.

LV

There is one principal and as it were radical distinction between different minds, in respect of philosophy and the sciences; which is this: that some minds are stronger and apter to mark the differences of things, others to mark their resemblances. The steady and acute mind can fix its contemplations and dwell and fasten on the subtlest distinctions: the lofty and discursive mind recognizes and puts together the finest and most general resemblances. Both kinds however easily err in excess, by catching the one at gradations the other at shadows.

LVI

There are found some minds given to an extreme admiration of antiquity, others to an extreme love and appetite for novelty; but few so duly tempered that they can hold the mean, neither carping at what has been well laid down by the ancients, nor despising what is well introduced by the moderns. This however turns to the great injury of the sciences and philosophy; since these affectations of antiquity and novelty are the humours of partisans rather than judgments; and truth is to be sought for not in the felicity of any age, which is an unstable thing, but in the light of nature and experience, which is eternal. These factions therefore must be abjured, and care must be taken that the intellect be not hurried by them into assent.

[8]*William* Gilbert (1544–1603). Gilbert wrote a treatise on magnetism and then went on to argue that the earth is a magnet. Bacon thought Gilbert's work was too narrow and one-dimensional. See Rees, 513.

LVII

Contemplations of nature and of bodies in their simple form break up and distract the understanding, while contemplations of nature and bodies[9] in their composition and configuration overpower and dissolve the understanding: a distinction well seen in the school of Leucippus and Democritus[10] as compared with the other philosophies. For that school is so busied with the particles that it hardly attends to the structure; while the others are so lost in admiration of the structure that they do not penetrate to the simplicity of nature. These kinds of contemplation should therefore be alternated and taken by turns; that so the understanding may be rendered at once penetrating and comprehensive, and the inconveniences above mentioned, with the idols which proceed from them, may be avoided.

LVIII

Let such then be our provision and contemplative prudence for keeping off and dislodging the *Idols of the Cave*, which grow for the most part either out of the predominance of a favorite subject, or out of an excessive tendency to compare or to distinguish, or out of partiality for particular ages, or out of the largeness or minuteness of the objects contemplated. And generally let every student of nature take this as a rule,—that whatever his mind seizes and dwells upon with peculiar satisfaction is to be held in suspicion, and that so much the more care is to be taken in dealing with such questions to keep the understanding even and clear.

---

[9]*By* "nature" and "bodies" Bacon refers here to all bodies and the whole of nature: looking just to the smallest parts or to the apparent wholes misses all the structures and motions in between.

[10]*See* note 7 above. Leucippus (Fifth Century B.C.E.) Pre-Socratic atomist philosopher; see Diogenes Laertius, *Lives of Eminent Philosophers*, IX: 30–33.

LIX

But the *Idols of the Market-place* are the most troublesome of all: idols which have crept into the understanding through the alliances of words and names. For men believe that their reason governs words; but it is also true that words react on the understanding; and this it is that has rendered philosophy and the sciences sophistical and inactive. Now words, being commonly framed and applied according to the capacity of the vulgar, follow those lines of division which are most obvious to the vulgar understanding. And whenever an understanding of greater acuteness or a more diligent observation would alter those lines to suit the true divisions of nature, words stand in the way and resist the change. Whence it comes to pass that the high and formal discussions of learned men end oftentimes in disputes about words and names; with which (according to the use and wisdom of the mathematicians) it would be more prudent to begin, and so by means of definitions reduce them to order. Yet even definitions cannot cure this evil in dealing with natural and material things; since the definitions themselves consist of words, and those words beget others: so that it is necessary to recur to individual instances, and those in due series and order; as I shall say presently when I come to the method and scheme for the formation of notions and axioms.

LX

The idols imposed by words on the understanding are of two kinds. They are either names of things which do not exist (for as there are things left unnamed through lack of observation, so likewise are there names which result from fantastic suppositions and to which nothing in reality corresponds), or they are names of things which exist, but yet confused and ill defined, and hastily and irregularly derived from realities. Of the former kind are Fortune, the Prime Mover, Planetary Orbits, Element of Fire, and like fictions which owe their origin to false and idle theories. And this class of idols is more easily expelled, because to get rid of them it is only necessary that all theories should be steadily rejected and dismissed as obsolete.

But the other class, which springs out of a faulty and unskillful abstraction, is intricate and deeply rooted. Let us take for example such a word as *humid*;[11] and see how far the several things which the word is used to signify agree with each other; and we shall find the word *humid* to be nothing else than a mark loosely and confusedly applied to denote a variety of actions which will not bear to be reduced to any constant meaning. For it both signifies that which easily spreads itself round any other body; and that which in itself is indeterminate and cannot solidise;[12] and that which readily yields in every direction; and that which easily divides and scatters itself; and that which easily unites and collects itself; and that which readily flows and is put in motion; and that which readily clings to another body and wets it; and that which is easily reduced to a liquid, or being solid easily melts. Accordingly when you come to apply the word,—if you take it in one sense, flame is humid; if in another, air is not humid; if in another, fine dust is humid; if in another, glass is humid. So that it is easy to see that the notion is taken by abstraction only from water and common and ordinary liquids, without any due verification.

There are however in words certain degrees of distortion and error. One of the least faulty kinds is that of names of substances, especially of lowest species and well-deduced (for the notion of *chalk* and of *mud* is good, of *earth* bad); a more faulty kind is that of actions, as *to generate, to corrupt, to alter*; the most faulty is of qualities (except such as are the immediate objects of the sense) as *heavy, light, rare, dense*, and the like. Yet in all these cases some notions are of necessity a little better than others, in proportion to the greater variety of subjects that fall within the range of the human sense.

---

[11]*Humidum*; see Rees, 514, who explains that the "term *humidum* in philosophical Latin can be used in the significances Bacon lists; the English translation masks this fact."

[12]*Remain* stable; see Rees, 94–95.

LXI

But the *Idols of the Theater* are not innate, nor do they steal into the understanding secretly, but are plainly impressed and received into the mind from the play-books of philosophical systems and the perverted rules of demonstration. To attempt refutations in this case would be merely inconsistent with what I have already said: for since we agree neither upon principles nor upon demonstrations there is no place for argument. And this is so far well, inasmuch as it leaves the honour of the ancients untouched. For they are no wise disparaged—the question between them and me being only as to the way. For as the saying is, the lame man who keeps the right road outstrips the runner who takes a wrong one. Nay it is obvious that when a man runs the wrong way, the more active and swift he is the further he will go astray.

But the course I propose for the discovery of sciences is such as leaves but little to the acuteness and strength of wits, but places all wits and understandings nearly on a level. For as in the drawing of a straight line or a perfect circle, much depends on the steadiness and practice of the hand, if it be done by aim of hand only, but if with the aid of rule or compass, little or nothing; so is it exactly with my plan. But though particular confutations would be of no avail, yet touching the sects and general divisions of such systems I must say something; something also touching the external signs which show that they are unsound; and finally something touching the causes of such great infelicity and of such lasting and general agreement in error; that so the access to truth may be made less difficult, and the human understanding may the more willingly submit to its purgation and dismiss its idols.

LXII

Idols of the Theater, or of Systems, are many, and there can be and perhaps will be yet many more. For were it not that now for many ages men's minds have been busied with religion and theology; and were it not that civil governments, especially monarchies, have been averse to such novelties, even in matters speculative; so that men labour therein to the peril and harming of their fortunes,—not only unrewarded, but exposed also to contempt and envy; doubtless there would have arisen many other philosophical sects like to those which in great variety flourished once among the Greeks. For as on the phenomena of the heavens many hypotheses may be constructed, so likewise (and more also) many various dogmas may be set up and established on the phenomena of philosophy. And in the plays of this philosophical theater you may observe the same thing which is found in the theater of the poets, that stories invented for the stage are more compact and elegant, and more as one would wish them to be, than true stories out of history.

In general however there is taken for the material of philosophy either a great deal out of a few things, or a very little out of many things; so that on both sides philosophy is based on too narrow a foundation of experiment and natural history, and decides on the authority of too few cases. For the Rational School of philosophers snatches from experience a variety of common instances, neither duly ascertained nor diligently examined and weighed, and leaves all the rest to meditation and agitation of wit.

There is also another class of philosophers, who having bestowed much diligent and careful labour on a few experiments, have thence made bold to educe and construct systems; wresting all other facts in a strange fashion to conformity therewith.

And there is yet a third class, consisting of those who out of faith and veneration mix their philosophy with theology and traditions; among whom the vanity of some has gone so far aside as to seek the origin of sciences among spirits and genii.[13] So that this parent stock of errors—this false philosophy—is of three kinds; the Sophistical, the Empirical, and the Superstitious.

---

[13]*Bacon* here refers to the scholastic science and magical writings of Heinrich Cornelius Agrippa (1468–1535): *De occulta philosophia.* See Rees, 514–15.

LXIII

The most conspicuous example of the first class was Aristotle, who corrupted natural philosophy by his logic: fashioning the world out of categories; assigning to the human soul, the noblest of substances, a genus from words of the second intention;[14] doing the business of density and rarity (which is to make bodies of greater or less dimensions, that is, occupy greater or less spaces), by the frigid distinction of act and power; asserting that single bodies have each a single and proper motion, and that if they participate in any other, then this results from an external cause; and imposing countless other arbitrary restrictions on the nature of things; being always more solicitous to provide an answer to the question and affirm something positive in words, than about the inner truth of things; a failing best shown when his philosophy is compared with other systems of note among the Greeks.[15] For the *homœomera* of Anaxagoras;[16] the Atoms of Leucippus and Democritus; the Heaven and Earth of Parmenides;[17] the Strife and Friendship of Empedocles;[18] Heraclitus's doctrine how bodies are resolved into the indifferent nature of fire, and remoulded into solids;[19] have all of them some taste of the natural philosopher,—some savour of the nature of things, and experience, and bodies; whereas in the physics of Aristotle you hear hardly anything but the words of logic; which in his metaphysics also, under a more imposing name, and more

[14]*Words* referring not to concrete things but rather to such things as the objects of thinking or logic.

[15]*For* a detailed discussion of Bacon's attack on Aristotle, see Rees, 515–17.

[16]*Homogeneous* atoms; see Diogenes Laertius, *Lives of Eminent Philosophers*, Anaxagoras (500–428 B.C.E.), II: 6–10.

[17]*Parmenides* (Fifth Century B.C.E.) formulated a cosmology with the earth at its center and a celestial region surrounding it. See Diogenes Laertius *Lives of Eminent Philosophers*, IX: 21–23.

[18]*Empedocles* (484–424 B.C.E.) held strife and friendship to be the eternal forces that move the four elements of earth, air, water, and fire.

[19]*See* note 3 above.

forsooth as a realist than a nominalist, he has handled over again.[20] Nor let any weight be given to the fact that in his books on animals and his problems, and other of his treatises, there is frequent dealing with experiments. For he had come to his conclusion before; he did not consult experience, as he should have done, in order to the framing of his decisions and axioms; but having first determined the question according to his will, he then resorts to experience, and bending her into conformity with his placets[21] leads her about like a captive in a procession; so that even on this count he is more guilty than his modern followers, the schoolmen, who have abandoned experience altogether.

LXIV

But the Empirical school of philosophy gives birth to dogmas more deformed and monstrous than the Sophistical or Rational school. For it has its foundations not in the light of common notions, (which though it be a faint and superficial light, is yet in a manner universal, and has reference to many things,) but in the narrowness and darkness of a few experiments. To those therefore who are daily busied with these experiments, and have infected their imagination with them, such a philosophy seems probable and all but certain; to all men else incredible and vain. Of this there is a notable instance in the alchemists and their dogmas; though it is hardly to be found elsewhere in these times, except perhaps in the philosophy of Gilbert. Nevertheless with regard to philosophies of this kind there is one caution not to be omitted; for I foresee that if ever men are roused by my admonitions to betake themselves seriously to experiment and bid farewell to sophistical doctrines, then indeed through the premature hurry of the understanding to leap or fly to universals and principles of things, great danger may be apprehended from philosophies of this kind; against which evil we ought even now to prepare.

[20]*See* Rees, 517.
[21]*Expressions* of assent.

LXV

But the corruption of philosophy by superstition and an admixture of theology is far more widely spread, and does the greatest harm, whether to entire systems or to their parts. For the human understanding is obnoxious to[22] the influence of the imagination no less than to the influence of common notions. For the contentious and sophistical kind of philosophy ensnares the understanding; but this kind, being fanciful and tumid and half poetical, misleads it more by flattery. For there is in man an ambition of the understanding, no less than of the will, especially in high and lofty spirits.

Of this kind we have among the Greeks a striking example in Pythagoras,[23] though he united with it a coarser and more cumbrous superstition; another in Plato and his school, more dangerous and subtle. It shows itself likewise in parts of other philosophies, in the introduction of abstract forms and final causes and first causes, with the omission in most cases of causes intermediate, and the like. Upon this point the greatest caution should be used. For nothing is so mischievous as the apotheosis of error; and it is a very plague of the understanding for vanity to become the object of veneration. Yet in this vanity some of the moderns have with extreme levity indulged so far as to attempt to found a system of natural philosophy on the first chapter of Genesis,[24] on the book of Job,[25] and other parts of the sacred writings; seeking for the dead among the living: which also makes the inhibition and repression of it the more important, because from this unwholesome mixture of things human and divine there arises not only a fantastic philosophy but also an

---

[22]*Exposed* to.

[23]*Bacon* thought that for Pythagoras (c. 582–500 B.C.E.) the world was made up of numbers. See *De augmentis scientiarium* 3: 6 and Rees, 518.

[24]*Bacon* refers here to the followers of Paracelsus (1493–1541). See Rees, 518–19 and *The Oxford Francis Bacon*, Vol. VI (1996), xliv–lii.

[25]*Bacon* refers here to the Coimbrans who used Job 37: 18 to refute the claim that the 1572 new star in Cassiopia was in fact a star at all. See Rees, 519.

heretical religion. Very meet[26] it is therefore that we be sober-minded, and give to faith that only which is faith's.[27]

LXVI

So much then for the mischievous authorities of systems, which are founded either on common notions, or on a few experiments, or on superstition. It remains to speak of the faulty subject-matter of contemplations, especially in natural philosophy. Now the human understanding is infected by the sight of what takes place in the mechanical arts, in which the alteration of bodies proceeds chiefly by composition or separation, and so imagines that something similar goes on in the universal nature of things. From this source has flowed the fiction of elements, and of their concourse for the formation of natural bodies. Again, when man contemplates nature working freely, he meets with different species of things, of animals, of plants, of minerals; whence he readily passes into the opinion that there are in nature certain primary forms which nature intends to educe, and that the remaining variety proceeds from hindrances and aberrations of nature in the fulfillment of her work, or from the collision of different species and the transplanting of one into another. To the first of these speculations we owe our primary qualities of the elements; to the other our occult properties and specific virtues; and both of them belong to those empty *compendia* of thought wherein the mind rests, and whereby it is diverted from more solid pursuits.[28] It is to better purpose that the physicians bestow their labour on the secondary qualities of matter, and the operations of attraction, repulsion, attenuation, conspissation,[29] dilatation, astriction, dissipation, maturation, and the like; and were it not that by

[26]*Proper.*
[27]*Matt.* 22: 21.
[28]*Bacon* is thinking here of scholastic science and magic as seen in the works of Heinrich Cornelius Agrippa (1468–1535) and Benito Pereira (1535–1610); see Rees, 573.
[29]*Thickening.*

those two compendia which I have mentioned (elementary qualities, to wit, and specific virtues) they corrupted their correct observations in these other matters,—either reducing them to first qualities and their subtle and incommensurable mixtures, or not following them out with greater and more diligent observation to third and fourth qualities, but breaking off the scrutiny prematurely,—they had made much greater progress. Nor are powers of this kind (I do not say the same, but similar) to be sought for only in the medicines of the human body, but also in the changes of all other bodies.

But it is a far greater evil that they make the quiescent principles, *wherefrom*, and not the moving principles, *whereby*, things are produced, the object of their contemplation and inquiry. For the former tend to discourse, the latter to works. Nor is there any value in those vulgar distinctions of motion which are observed in the received system of natural philosophy, as generation, corruption, augmentation, diminution, alteration, and local motion. What they mean no doubt is this:—If a body, in other respects not changed, be moved from its place, this is *local motion*; if without change of place or essence, it be changed in quality, this is *alteration*; if by reason of the change the mass and quantity of the body do not remain the same, this is *augmentation* or *diminution*; if they be changed to such a degree that they change their very essence and substance and turn to something else, this is *generation* and *corruption*. But all this is merely popular, and does not at all go deep into nature; for these are only measures and limits, not kinds of motion. What they intimate is *how far*, not *by what means*, or *from what source*. For they do not suggest anything with regard either to the desires of bodies or to the development of their parts: it is only when that motion presents the thing grossly and palpably to the sense as different from what it was, that they begin to mark the division. Even when they wish to suggest something with regard to the causes of motion, and to establish a division with reference to them, they introduce with the greatest negligence a distinction between motion natural and violent; a distinction which is itself drawn entirely from a vulgar notion, since all violent motion is also in fact natural; the external efficient simply setting nature working otherwise than it was before. But if, leaving all this, anyone shall observe (for instance) that there is in bodies a desire of mutual contact, so as not to suffer the unity of nature to be quite separated or broken and a vacuum thus made; or if anyone say that there is in bodies a desire of resuming their natural dimensions or tension, so that if compressed within or extended beyond them, they immediately strive to recover themselves, and fall back to their old volume and extent; or if anyone say that there is in bodies a desire of congregating toward masses of kindred nature,—of dense bodies, for instance, towards the globe of the earth, of thin and rare bodies towards the compass of the sky; all these and the like are truly physical kinds of motion;—but those others are entirely logical and scholastic, as is abundantly manifest from this comparison.

Nor again is it a less evil, that in their philosophies and contemplations their labour is spent in investigating and handling the first principles of things and the highest generalities of nature; whereas utility and the means of working result entirely from things intermediate. Hence it is that men cease not from abstracting nature till they come to potential and uninformed matter, nor on the other hand from dissecting nature till they reach the atom; things which, even if true, can do but little for the welfare of mankind.

LXVII

A caution must also be given to the understanding against the intemperance which systems of philosophy manifest in giving or withholding assent; because intemperance of this kind seems to establish Idols and in some sort to perpetuate them, leaving no way open to reach and dislodge them.

This excess is of two kinds: the first being manifest in those who are ready in deciding, and render sciences dogmatic and magisterial; the other in those who deny that we can know anything, and so introduce a wandering kind of inquiry that leads to nothing; of which kinds the former subdues, the latter weakens the understanding. For the philosophy of Aristotle, after having by hostile confutations destroyed all the rest (as the Ottomans serve their brothers[30]), has laid down the law on all points; which done, he proceeds himself to raise new questions of his own suggestion, and dispose of them likewise; so that nothing may remain that is not certain and decided: a practice which holds and is in use among his successors.

The school of Plato, on the other hand, introduced *Acatalepsia*,[31] at first in jest and irony, and in disdain of the older sophists, Protagoras,[32] Hippias,[33] and the rest, who were of nothing else so much ashamed as of seeming to doubt about anything. But the New Academy[34] made a dogma of it, and held it as a tenet. And though their's is a fairer seeming way than arbitrary decisions; since they say that they by no means destroy all investigation, like Pyrrho and his Refrainers,[35] but allow of some things to be followed as probable, though of none to be maintained as true; yet still when the human mind has once despaired of finding truth, its interest in all things grows fainter; and the result is that men turn aside to pleasant disputations and discourses and roam as it were from object to object, rather than keep on a course of severe inquisition. But, as I said at the beginning and am ever urging, the human senses and understanding, weak as they are, are not to be deprived of their authority, but to be supplied with helps.

---

[30]*Bacon* refers here to the Ottoman practice of fratricide, whereby a new Sultan would have his brothers and half-brothers strangled.

[31]*Doubt* that anything can be known for certain.

[32]*See* note 2 above.

[33]*Greek* Sophist and contemporary of Socrates.

[34]*According* to Diogenes Laertius the New Academy in Athens was founded by Lacydes in the year July 241–June 240 B.C.E. See *Lives of Eminent Philosophers* IV: 59–61.

[35]*"Refrainers"* is Spedding's translation of Bacon's term *Ephectics*. According to Diogenes Laertius, Pyrrho was an extreme skeptic who lived from c. 360–270 B.C.E. His students were called *ephetics* because of their mind-frame of suspended judgment after their inquiries. *Lives of Eminent Philosophers*, IX: 61–70.

LXVIII

So much concerning the several classes of Idols, and their equipage: all of which must be renounced and put away with a fixed and solemn determination, and the understanding thoroughly freed and cleansed; the entrance into the kingdom of man, founded on the sciences, being not much other than the entrance into the kingdom of heaven, where-into none may enter except as a little child.[36]

Synopsis

According to Bacon, the first task of his "great instauration" was to recognize that human reason, bestowed by our Creator, is not such a wonderful gift. Reason may be the unique possession of the human animal, but so too are self-delusion, prejudice, mendacity, and superstition. For knowledge to progress, the diseases of reason must be diagnosed and cured so far as possible. Bacon presented his diagnosis in his doctrine called "The Idols of the Mind," presented in aphorisms 38 through 68 (above) of the first book of his *Novum organum*. Bacon outlined four categories of defects to which unaided reason is inclined, defects that lead the "rational animal" to false, distorted, fruitless, and superstitious conceptions of the natural world.

The first category he called the "idols of the tribe." These were entirely intrinsic to reason: The mind's propensity to imagine greater order and regularity in the world than truly exists; its habit of falling in love with an opinion and ignoring all evidence to the contrary; its preference for affirmative instances over negative ones; its propensity to be impressed by sudden and simultaneous occurrences; its restless propensity to seek a cause behind causes and so to dream up things like a limitless world or the paradoxical idea of past and future infinities or, worst of all, the world's "final cause"; its tendency to be influenced by what we want and feel; the weakness of its

[36]*Mark* 10: 15.

senses and inclination to be satisfied with what it sees; and its propensity for abstraction and to think that abstractions are really things.

The second category is the "idols of the cave." These are the various aspects of character, education, and habit of each individual mind. Some people become obsessed with a single science and apply it then to everything, as Bacon says Aristotle did with his logic and Gilbert did with his loadstone. Some people are more inclined to see the similarities of things and others to see the differences. Some people fall in love with antiquity and others are zealous for novelty. Some are obsessed with small particles and others with bigger structures: the best thing is to combine both points of view.

The third category is the "idols of the marketplace." These idols are the "most troublesome," says Bacon, and consist in the distortions of the understanding imposed by words and names. There are two kinds: names of things that don't really exist, such as "fortune," "element of fire," and "prime mover," and names of things that are hastily abstracted from natural actions that really have nothing in common, such as using the term "humid" to mean something that can cover over something else, and also something that "readily yields in in every direction."

The last category, the "idols of the theatre," is neither innate nor hidden. These idols result from the historical weight of "the playbooks of philosophical systems" and "perverted rules of demonstration." As regards them, Bacon says that there is no point in refuting them since his new principles for the discovery of sciences have nothing in common with them, and in any event his new principles take the place of differences of wit among men. So Bacon rather comments more generally on the three kinds of errors that strut about on the philosophical stage: "the Sophistical" (or "Rational"), "the Empirical," and "the Superstitious."

Aristotle is the exemplar of the sophistical error, since he corrupted his natural science with his logic and "categories" and all kinds of airy abstractions rather than focusing on the "inner truth of things." Even when he engaged in experiments, says Bacon, he did so with the conclusion he wanted in mind, and in this respect he was worse than his modern scholastic followers who at least abandon experience altogether and so do not disgrace it. As bad as Aristotle was, the Empirical school is even worse. At least Aristotle began with common notions, but the Empirical school begins with the "narrowness and darkness of a few experiments." This isn't widespread in his time says Bacon, although we see it in the dogmas of the alchemists and the philosophy of Gilbert. Bacon worries especially about such monomaniacal nonsense because it bears some slight resemblance to his own reliance on experience and experiment and thus always stands ready to derail natural science properly understood.

Superstitious philosophy is more widely found throughout history and is exemplified by the thought of Pythagoras and Plato among the Greeks, wherein the influence of superstition and religion led to the corruption of natural philosophy with "abstract forms and final causes and first causes," all of which rather flatter the ambitions of the understanding than discover the real actions of nature. In modern times the same corruption can be seen in thinkers who try to model natural science on the Bible and get only fantastic and heretical results.

Bacon ends his discussion of the Idols of the Mind by turning from the "mischievous authorities" of systems to the bad subject matter of the accepted natural philosophy. The first source of error in this regard is that the human mind looks at the natural world from the perspective of the mechanical arts, in which things are produced either by composition or separation. This leads to the false idea that nature works by such composition and separation of "elements" (such as fire, air, water, and earth). Another error springs from the mind's jumping from the observation of nature working freely to the belief that some of the species of animals, plants, and minerals are nature's "primary" aim and everything else is the result of obstacles in the way of nature or aberrations or is derived from the "collision" of different species or the transplanting of one into another. This leads to the attribution of "occult properties" and "specific virtues," unknowable by reason, to things that are not "primary." All this nonsense is just the resting of the mind and avoidance of more productive investigation. Even when physicians seeking medicines did spend time studying the qualities and operations of such things as attraction, repulsion, attenuation, thickening, dilation, and maturation, they corrupted their otherwise correct observations by the nonsensical dichotomy of elementary qualities and specific virtues.

But the greater evil is that the practitioners of the received natural philosophy look to quiescent principles and ask what comes from what, rather looking to moving principles and asking about how or "whereby" things are produced. And what they do say about motions is completely useless: all they address is the measure and limits of motions, but they ignore the kinds of motions that occur in nature by not asking "by what means or from what source" does a motion occur? Bacon ends by noting that if you consider just some genuinely physical kinds of motions it becomes obvious that the traditionalists are simply logical and scholastic. It is equally bad that they look for first principles and greatest generalities about nature rather than investigating intermediate causes: even if one found the very first and unformed matter of nature, or dissected that matter until finding its very smallest part, that would do no practical good for humanity.

In the penultimate aphorism of the Idols of the Mind, Bacon warns against the intemperance by which philosophical systems give or withhold "assent." There are two opposing extremes: some philosophers are too ready to decide things and are dogmatic, while others deny that we can know anything. The first subdues the mind and the latter weakens it. Aristotle, "as do the Ottomans serve their brothers," destroyed everything that came before him and laid down the philosophical law "on all points," and then questioned those points on his own and then disposed of his doubts about them as well.

Plato, on the other hand, introduced "actalepsia" (doubt that anything can be known for certain) at first as a joke to prod the "older sophists" such as Protagoras and Hippias, who were ashamed to seem to have doubts about anything. But

Plato's later successors in the New Academy made a dogma of this skepticism. Bacon comments that the philosophers of the New Academy were better than "Pyrrho and his Refrainers" because while the latter denied the possibility of any knowledge, the New Academicians allowed for probability. But Bacon notes that once the possibility of truth is forsaken, the mind's interest in things wanes and flits from object to object with no power to engage in "severe inquisition." The human senses and understanding, says Bacon, however weak they may be, need help, not depressed authority.

As if to repair to his theme of a new beginning, Bacon closes with a twist on the Gospel of Mark. We must be cleansed of the Idols and become as children if we are to enter the Kingdom of man founded on the sciences.

Questions

Do you think any or all of Bacon's Idols of the Mind affect your own thinking or understanding of the world?

What is the implication for religious faith of Bacon's doctrine of the Idols of the Mind?

Does Bacon think that the Idols of the Mind can be completely expunged from human reason?

NEW ATLANTIS:

A WORK UNFINISHED.

WRITTEN BY

THE RIGHT HONOURABLE

FRANCIS LORD VERULAM, VISCOUNT ST. ALBAN[1]

---

[1]*Bacon's* titles. He was also knighted.

TO THE READER

This fable my Lord devised, to the end that he might exhibit therein a model or description of a college instituted for the interpreting of nature and the producing of great and marvellous works for the benefit of men, under the name of Salomon's House, or the College of the Six Days' Works. And even so far his Lordship hath proceeded, as to finish that part. Certainly the model is more vast and high than can possibly be imitated in all things; notwithstanding most things therein are within men's power to effect. His Lordship thought also in this present fable to have composed a frame of Laws, or of the best state or mould of a commonwealth; but foreseeing it would be a long work, his desire of collecting the Natural History diverted him, which he preferred many degrees before it.

This work of the *New Atlantis* (as much as concerneth the English edition) his Lordship designed for this place;[2] in regard it hath so near affinity (in one part of it) with the preceding Natural History.

W. RAWLEY[3]

---

[2] *The New Atlantis* was published in 1627, after Bacon's death. It appeared at the end of the volume containing the *Sylva sylvarum* ("A Forest of Materials"), a part of his largely unfinished natural history.
[3] *Bacon's* secretary.

NEW ATLANTIS

We sailed from Peru, (where we had continued[4] by the space
of one whole year,) for China and Japan, by the South Sea;[5]
taking with us victuals for twelve months; and had good winds
from the east, though soft and weak, for five months' space and
more. But then the wind came about,[6] and settled in the west
for many days, so as we could make little or no way, and were
sometimes in purpose[7] to turn back. But then again there arose
strong and great winds from the south, with a point east; which
carried us up (for all that we could do) towards the north: by
which time our victuals failed us, though we had made good
spare of them.[8] So that finding ourselves in the midst of the
greatest wilderness of waters in the world, without victual, we
gave ourselves for lost men, and prepared for death. Yet we did
lift up our hearts and voices to God above, who *showeth his
wonders in the deep*;[9] beseeching him of his mercy, that as in
the beginning he discovered[10] the face of the deep, and brought
forth dry land, so he would now discover land to us, that we
might not perish. And it came to pass that the next day about
evening, we saw within a kenning[11] before us, towards the
north, as it were thick clouds, which did put us in some hope of
land; knowing how that part of the South Sea was utterly
unknown; and might have islands or continents, that hitherto

---

[4]*Stayed* for.
[5]*Pacific* Ocean.
[6]*Came* from the opposite direction.
[7]*Decided*, resolved.
[8]*Conserved* them.
[9]*Psalms* 107: 23–32.
[10]*Uncovered*, drew back.
[11]*Distance* measured by the range of sight: 20 miles.

New Atlantis and The Great Instauration, Second Edition.
Edited by Jerry Weinberger.
© 2017 John Wiley & Sons, Inc. Published 2017 by John Wiley & Sons, Inc.

were not come to light. Wherefore we bent[12] our course thither, where we saw the appearance of land, all that night; and in the dawning of the next day, we might plainly discern that it was a land, flat to our sight, and full of boscage;[13] which made it shew the more dark. And after an hour and a half's sailing, we entered into a good haven, being the port of a fair city; not great indeed, but well built, and that gave a pleasant view from the sea: and we thinking every minute long till we were on land, came close to the shore, and offered[14] to land. But straightways we saw divers[15] of the people, with bastons[16] in their hands, as it were forbidding us to land; yet without any cries or fierceness, but only as warning us off by signs that they made. Whereupon being not a little discomforted, we were advising with ourselves what we should do. During which time there made forth to us a small boat, with about eight persons in it; whereof one of them had in his hand a tipstaff[17] of a yellow cane, tipped at both ends with blue, who came aboard our ship, without any show of distrust at all. And when he saw one of our number present himself somewhat afore the rest, he drew forth a little scroll of parchment, (somewhat yellower than our parchment, and shining like the leaves of writing tables,[18] but otherwise soft and flexible,) and delivered it to our foremost man. In which scroll were written in ancient Hebrew, and in ancient Greek, and in good Latin of the School,[19] and in Spanish, these words; "Land ye not, none[20] of you; and provide to be gone from this coast within sixteen days, except you have further time given you. Meanwhile, if you want fresh water, or victual, or help for your sick, or that your ship

[12]*Turned.*

[13]*Forest.*

[14]*Showed* intention to.

[15]*Various* (not people who dive!).

[16]*Batons*, truncheons.

[17]*A* staff with a tip or cap of metal carried as a badge by certain officials.

[18]*Tablets* for inscriptions.

[19]*Latin* as written by the Scholastic philosophers, university Latin.

[20]*An* acceptable double negative in Bacon's time.

needeth repair, write down your wants, and you shall have that which belongeth to mercy." This scroll was signed with a stamp of cherubins'[21] wings, not spread but hanging downwards, and by them a cross. This being delivered, the officer returned, and left only a servant with us to receive our answer. Consulting hereupon amongst ourselves, we were much perplexed. The denial of landing and hasty warning us away troubled us much; on the other side, to find that the people had languages[22] and were so full of humanity, did comfort us not a little. And above all, the sign of the cross to that instrument[23] was to us a great rejoicing, and as it were a certain presage of good. Our answer was in the Spanish tongue; "That for our ship, it was well; for we had rather met with calms and contrary winds than any tempests. For our sick, they were many, and in very ill case;[24] so that if they were not permitted to land, they ran danger of their lives." Our other wants we set down in particular; adding, "that we had some little store of merchandise, which if it pleased them to deal for, it might supply our wants without being chargeable unto them."[25] We offered some reward in pistolets[26] unto the servant, and a piece of crimson velvet to be presented to the officer; but the servant took them not, nor would scarce look upon them; and so left us, and went back in another little boat which was sent for him.

About three hours after we had dispatched our answer, there came towards us a person (as it seemed) of place.[27] He had on him a gown with wide sleeves, of a kind of water chamolet,[28] of an excellent azure colour, far more glossy than ours; his under apparel was green; and so was his hat, being in the form of a

[21]*A* biblical figure with wings, a human head, and an animal body.
[22]*Were* conversant in European languages.
[23]*The* scroll.
[24]*Poor* condition.
[25]*Without* costing them anything.
[26]*Gold* coins.
[27]*High* rank.
[28]*A* costly oriental fabric.

turban, daintily made, and not so huge as the Turkish turbans; and the locks of his hair came down below the brims of it. A reverend man was he to behold. He came in a boat, gilt in some part of it, with four persons more only in that boat; and was followed by another boat, wherein were some twenty. When he was come within a flight-shot[29] of our ship, signs were made to us that we should send forth some to meet him upon the water; which we presently did in our ship-boat,[30] sending the principal man amongst us save one,[31] and four of our number with him. When we were come within six yards of their boat, they called to us to stay,[32] and not to approach farther; which we did. And thereupon the man whom I before described stood up, and with a loud voice in Spanish, asked, "Are ye Christians?" We answered, "We were;" fearing the less, because of the cross we had seen in the subscription. At which answer the said person lifted up his right hand towards heaven, and drew it softly to his mouth, (which is the gesture they use when they thank God,) and then said: "If ye will swear (all of you) by the merits of the Saviour that ye are no pirates, nor have shed blood lawfully nor unlawfully within forty days past,[33] you may have licence to come on land." We said, "We were all ready to take that oath." Whereupon one of those that were with him, being (as it seemed) a notary, made an entry of this act. Which done, another of the attendants of the great person, which was with him in the same boat, after his lord had spoken a little to him, said aloud; "My lord would have you know, that it is not of pride or greatness that he cometh not aboard your ship; but for that in your answer you declare that you have many sick amongst you, he was warned by the Conservator of Health of the city that he should keep a distance." We bowed ourselves towards him, and answered, "We were his humble servants;

---

[29]*The* flight distance of a long-range arrow.

[30]*Small* boat carried or towed by a ship.

[31]*Second* in command.

[32]*Stop.*

[33]*See* Exodus 20: 13; 21: 12–14; Numbers 35: 9–34, Deut. 5: 17; cf. Maimonides, *Guide of the Perplexed*, III, 40, 41.

and accounted for great honour and singular humanity
towards us that which was already done; but hoped well
that the nature of the sickness of our men was not infectious."
So he returned; and a while after came the notary to us aboard
our ship; holding in his hand a fruit of that country, like an
orange, but of color between orange-tawney and scarlet, which
cast a most excellent odour. He used it (as it seemeth) for a
preservative against infection. He gave us our oath; "By the
name of Jesus and his merits:" and after told us that the next
day by six of the clock in the morning we should be sent to, and
brought to the Strangers' House, (so he called it,) where we
should be accomodated of things both for our whole[34] and for
our sick. So he left us; and when we offered him some pistolets,
he smiling said, "He must not be twice paid for one labour:"
meaning (as I take it) that he had salary sufficient of the state
for his service. For (as I after learned) they call an officer that
taketh rewards, *twice paid*.

The next morning early, there came to us the same officer
that came to us at first with his cane, and told us, "He came to
conduct us to the Strangers' House; and that he had prevented
the hour,[35] because we might have the whole day before us for
our business. "For," said he, "if you will follow my advice,
there shall first go with me some few of you, and see the place,
and how it may be made convenient for you; and then you may
send for your sick, and the rest of your number which ye will
bring on land." We thanked him, and said, "That this care
which he took of desolate strangers God would reward." And
so six of us went on land with him: and when we were on land,
he went before us, and turned to us, and said, "He was but our
servant, and our guide." He led us through three fair streets;
and all the way we went there were gathered some people on
both sides standing in a row; but in so civil a fashion, as if it had
been not to wonder at us but to welcome us; and divers of them,
as we passed by them, put their arms a little abroad; which is
their gesture when they bid any welcome. The Strangers' House

[34]*Well.*
[35]*He* had come early.

is a fair and spacious house, built of brick, of somewhat a bluer colour than our brick; and with handsome windows, some of glass, some of a kind of cambric[36] oiled. He brought us first into a fair parlour above stairs, and then asked us, "What number of persons we were? And how many sick?" We answered, "We were in all (sick and whole) one and fifty persons, whereof our sick were seventeen." He desired us to have patience a little, and to stay till he came back to us; which was about an hour after; and then he led us to see the chambers which were provided for us, being in number nineteen: they having cast it[37] (as it seemeth) that four of those chambers, which were better than the rest, might receive four of the principal men of our company, and lodge them alone by themselves; and the other fifteen chambers were to lodge us two and two together.[38] The chambers were handsome and cheerful chambers, and furnished civilly. Then he led us to a long gallery, like a dorture,[39] where he showed us all along the one side (for the other side was but wall and window) seventeen cells, very neat ones, having partitions of cedar wood. Which gallery and cells, being in all forty, (many more than we needed,) were instituted as an infirmary for sick persons. And he told us withal,[40] that as any of our sick waxed[41] well, he might be removed from his cell to a chamber; for which purpose there were set forth ten spare chambers, besides the number we spake of before. This done, he brought us back to the parlour, and lifting up his cane a little, (as they do when they give any charge or command,) said to us, "Ye are to know that the custom of the land requireth, that after this day and to-morrow, (which we give you for removing of your people from your ship,) you are to keep within doors for three days. But let it not trouble you, nor do not think yourselves

[36]*A kind of white linen.*
[37]*Reckoned.*
[38]*By twos.*
[39]*Dormitory.*
[40]*Moreover.*
[41]*Became.*

restrained, but rather left to your rest and ease. You shall want nothing, and there are six of our people appointed to attend you, for any business you may have abroad." We gave him thanks with all affection and respect, and said, "God surely is manifested in this land." We offered him also twenty pistolets; but he smiled, and only said; "What? twice paid!" And so he left us. Soon after our dinner was served in; which was right good viands,[42] both for bread and meat: better than any collegiate diet[43] that I have known in Europe. We had also drink of three sorts, all wholesome and good; wine of the grape; a drink of grain, such as is with us our ale, but more clear; and a kind of cider made of a fruit of that country; a wonderful pleasing and refreshing drink. Besides, there were brought in to us great store of those scarlet oranges for our sick; which (they said) were an assured remedy for sickness taken at sea. There was given us also a box of small grey or whitish pills, which they wished our sick should take, one of the pills every night before sleep; which (they said) would hasten their recovery. The next day, after that our trouble of carriage and removing of our men and goods out of our ship was somewhat settled and quiet, I thought good to call our company together; and when they were assembled said unto them; "My dear friends, let us know ourselves, and how it standeth with us. We are men cast on land, as Jonas[44] was out of the whale's belly, when we were as buried in the deep: and now we are on land, we are but between death and life; for we are beyond both the old world and the new; and whether ever we shall see Europe, God only knoweth. It is a kind of miracle hath brought us hither: and it must be little less that shall bring us hence.[45] Therefore in regard of our deliverance past, and our danger present and to come, let us look up to God, and every man reform his own ways. Besides we are come here amongst a Christian people, full of piety and humanity: let us not bring

---

[42]*Food.*
[43]*Food* served for an organized society.
[44]*Jonah.* See Jonah 1–4.
[45]*From* this place.

that confusion of face[46] upon ourselves, as to show our vices or unworthiness before them. Yet there is more. For they have by commandment (though in form of courtesy) cloistered us within these walls for three days: who knoweth whether it be not to take some taste[47] of our manners and conditions? and if they find them bad, to banish us straightways; if good, to give us further time. For these men that they have given us for attendance may withal have an eye upon us. Therefore for God's love, and as we love the weal[48] of our souls and bodies, let us so behave ourselves as we may be at peace with God, and may find grace in the eyes of this people." Our company with one voice thanked me for my good admonition, and promised me to live soberly and civilly, and without giving any the least occasion of offence. So we spent our three days joyfully and without care, in expectation what would be done with us when they were expired. During which time, we had every hour joy of the amendment[49] of our sick; who thought themselves cast into some divine pool of healing,[50] they mended so kindly[51] and so fast.

The morrow[52] after our three days were past, there came to us a new man that we had not seen before, clothed in blue as the former was, save that his turban was white, with a small red cross on the top. He had also a tippet[53] of fine linen. At his coming in, he did bend to us a little, and put his arms abroad. We of our parts saluted him in a very lowly and submissive manner; as looking that from him we should receive sentence of life or death. He desired to speak with some few of us: whereupon six of us only stayed, and the rest avoided[54] the room.

[46]*Shame.*
[47]*To* observe.
[48]*Well-being.*
[49]*Recovery.*
[50]*Cf.* John 5: 2–9.
[51]*Naturally.*
[52]*Morning.*
[53]*Long* scarf or cape.
[54]*Left.*

He said, "I am by office governor of this House of Strangers, and by vocation I am a Christian priest; and therefore am come to you to offer you my service, both as strangers and chiefly as Christians. Some things I may tell you, which I think you will not be unwilling to hear. The state hath given you licence to stay on land for the space of six weeks: and let it not trouble you if your occasions ask[55] further time, for the law in this point is not precise; and I do not doubt but myself shall be able to obtain for you such further time as may be convenient. Ye shall also understand, that the Strangers' House is at this time rich, and much aforehand;[56] for it hath laid up revenue these thirty-seven years; for so long it is since any stranger arrived in this part: and therefore take ye no care; the state will defray you[57] all the time you stay; neither shall you stay one day the less for that. As for any merchandise ye have brought, ye shall be well used,[58] and have your return either in merchandise or in gold and silver: for to us it is all one. And if you have any other request to make, hide it not. For ye shall find we will not make your countenance to fall[59] by the answer ye shall receive. Only this I must tell you, that none of you must go above a *karan*"[60] (that is with them a mile and an half) "from the walls of the city, without especial leave." We answered, after we had looked awhile one upon another, admiring this gracious and parent-like usage;[61] "That we could not tell what to say: for we wanted[62] words to express our thanks; and his noble free offers left us nothing to ask. It seemed to us that we had before us a

[55]*Your* needs require.
[56]*Prepared* or provided for the future.
[57]*Pay* your expenses.
[58]*Well* treated, treated fairly.
[59]*We* will not disappoint you.
[60]*This* comes from the Hebrew word *keren*, which means "horn" and is used to symbolize people's strength. It is used to refer to a ruler of the Davidic line; see Psalms 132: 17; Ezekiel 29: 21. The verb form means to send out rays or display horns.
[61]*Treatment.*
[62]*Lacked.*

picture of our salvation in heaven; for we that were awhile since[63] in the jaws of death, were now brought into a place where we found nothing but consolations. For the commandment laid upon us, we would not fail to obey it, though it was impossible but our hearts should be inflamed to tread further upon this happy and holy ground." We added; "That our tongues should first cleave to the roofs of our mouths,[64] ere we should forget either his reverend person or this whole nation in our prayers." We also most humbly besought him to accept of us as his true servants, by as just a right as ever men on earth were bounden; laying and presenting both our persons and all we had at his feet. He said; "He was a priest, and looked for a priest's reward: which was our brotherly love and the good of our souls and bodies." So he went from us, not without tears of tenderness in his eyes; and left us also confused[65] with joy and kindness, saying amongst ourselves, "That we were come into a land of angels, which did appear to us daily and prevent us[66] with comforts, which we thought not of, much less expected."

The next day, about ten of the clock, the governor came to us again, and after salutations said familiarly, "That he was come to visit us": and called for a chair, and sat him down: and we, being some ten of us, (the rest were of the meaner sort,[67] or else gone abroad,) sat down with him. And when we were set, he began thus: "We of this island of Bensalem,"[68] (for so they call it in their language,) "have this; that by means of our solitary situation, and of the laws of secrecy which we have for our travellers, and our rare admission of strangers, we know well most part of the habitable world, and are ourselves unknown. Therefore because he that knoweth least is fittest to ask

---

[63]*Just* before.
[64]*Psalms* 137: 6.
[65]*Amazed*.
[66]*Anticipate* our needs.
[67]*Of* a humbler class.
[68]"*Bensalem*" is a combination of Hebrew words (*ben, shalem*) meaning "son or offspring of peace, safety, and completeness."

questions, it is more reason, for the entertainment of the time,[69] that ye ask me questions, than that I ask you." We answered; "That we humbly thanked him that he would give us leave so to do: and that we conceived by the taste we had already, that there was no worldly thing on earth more worthy to be known than the state of that happy land. But above all," (we said,) "since that we were met from the several ends of the world, and hoped assuredly that we should meet one day in the kingdom of heaven, (for that we were both parts Christians,) we desired to know (in respect that land was so remote, and so divided by vast and unknown seas, from the land where our Saviour walked on earth,) who was the apostle[70] of that nation, and how it was converted to the faith?" It appeared in his face that he took great contentment in this our question: he said, "Ye knit my heart to you, by asking this question in the first place; for it sheweth that you *first seek the kingdom of heaven*;[71] and I shall gladly and briefly satisfy your demand.

"About twenty years after the ascension of our Saviour, it came to pass that there was seen by the people of Renfusa,[72] (a city upon the eastern coast of our island,)[73] within night, (the night was cloudy and calm,) as it might be some mile into the sea,[74] a great pillar of light; not sharp, but in form of a column or cylinder, rising from the sea a great way up towards heaven: and on the top of it was seen a large cross of light, more bright and resplendent than the body of the pillar. Upon which so strange a spectacle, the people of the city gathered apace[75] together upon the sands, to wonder; and so after put themselves into a number of small boats, to go nearer to this

---

[69]*For* passing the time.
[70]*The* missionary who first plants Christianity in a region.
[71]*Matt.* 6: 33.
[72]*"Renfusa"* is a combination of the Greek words *rhen* and *phusis* meaning "sheep-natured" or "sheep-like."
[73]*Since* the wind bringing the sailors to the island was from the south with "a point east," the sailors landed on the east side of the island.
[74]*A* mile out to sea.
[75]*Quickly.*

marvellous sight. But when the boats were come within about sixty yards of the pillar, they found themselves all bound, and could go no further; yet so as they might move to go about,[76] but might not approach nearer: so as the boats stood all as in a theatre, beholding this light as an heavenly sign. It so fell out,[77] that there was in one of the boats one of the wise men of the society of Salomon's House; which house or college (my good brethren) is the very eye[78] of this kingdom; who having awhile attentively and devoutly viewed and contemplated this pillar and cross, fell down upon his face; and then raised himself upon his knees, and lifting up his hands to heaven, made his prayers in this manner:

" 'Lord God of heaven and earth, thou hast vouchsafed of thy grace to those of our order, to know thy works of creation, and the secrets of them; and to discern (as far as appertaineth to the generations of men) between divine miracles, works of nature, works of art, and impostures and illusions of all sorts. I do here acknowledge and testify before this people, that the thing which we now see before our eyes is thy Finger and a true Miracle; and forasmuch as[79] we learn in our books that thou never workest miracles but to a divine and excellent end, (for the laws of nature are thine own laws, and thou exceedest them not but upon great cause,) we most humbly beseech thee to prosper[80] this great sign, and to give us the interpretation and use of it in mercy; which thou dost in some part secretly promise by sending it unto us.' "

"When he had made his prayer, he presently found the boat he was in moveable and unbound; whereas all the rest remained still fast; and taking that for an assurance of leave to approach, he caused the boat to be softly and with silence rowed towards the pillar. But ere he came near it, the pillar and cross of light brake up, and cast itself abroad, as it were, into a

---

[76]*To* turn around to go back.
[77]*It* so happened.
[78]*Seat* of intelligence.
[79]*Since*.
[80]*To* cause to succeed.

firmament of many stars; which also vanished soon after, and there was nothing left to be seen but a small ark or chest of cedar, dry, and not wet at all with water, though it swam. And in the fore-end of it, which was towards him, grew a small green branch of palm; and when the wise man had taken it with all reverence into his boat, it opened of itself, and there were found in it a Book and a Letter; both written in fine parchment, and wrapped in sindons of linen.[81] The Book contained all the canonical books[82] of the Old and New Testament, according as you have them, (for we know well what the Churches with you receive); and the Apocalypse[83] itself, and some other books of the New Testament which were not at that time written,[84] were nevertheless in the Book. And for the Letter, it was in these words:

" 'I Bartholomew,[85] a servant of the Highest, and Apostle of Jesus Christ, was warned by an angel that appeared to me in a vision of glory, that I should commit this ark to the floods of the sea. Therefore I do testify and declare unto that people where God shall ordain this ark to come to land, that in the same day is come unto them salvation and peace and goodwill, from the Father, and from the Lord Jesus.' "

"There was also in both these writings, as well the Book as the Letter, wrought a great miracle, conform to that of the Apostles in the original Gift of Tongues.[86] For there being at that time in this land Hebrews, Persians, and Indians, besides the natives, every one read upon the Book and Letter, as if they had been written in his own language. And thus was this land saved from infidelity (as the remain of the old world was from

[81]*Wrappings* of fine, thin linen.
[82]*The* canonical books of the Bible are those accepted as Holy Scripture. The Apocrypha are those books included in the Septuagint (Greek Bible) and the Vulgate (Latin Bible) but excluded from the Jewish and Protestant canons of the Old Testament.
[83]*The* Revelation to John.
[84]*At* least Acts, Paul's Epistles, and the Revelation to John!
[85]*St*. Bartholomew was the man said to have preached the gospel to India; *i.e.*, the islands off of the southern coast of Asia.
[86]*Acts* 2: 1–16.

water) by an ark, through the apostolical and miraculous evangelism of St. Bartholomew." And here he paused, and a messenger came, and called him from us. So this was all that passed in that conference.

The next day, the same governor came again to us immediately after dinner, and excused himself, saying, "That the day before he was called from us somewhat abruptly, but now he would make us amends, and spend time with us, if we held his company and conference agreeable." We answered, "That we held it so agreeable and pleasing to us, as we forgot both dangers past and fears to come, for the time we heard him speak; and that we thought an hour spent with him, was worth years of our former life." He bowed himself a little to us, and after we were set again, he said; "Well, the questions are on your part." One of our number said, after a little pause; "That there was a matter we were no less desirous to know, than fearful to ask, lest we might presume too far. But encouraged by his rare[87] humanity towards us, (that could scarce think ourselves strangers, being his vowed and professed servants,) we would take the hardiness[88] to propound it: humbly beseeching him, if he thought it not fit to be answered, that he would pardon it, though he rejected it." We said; "We well observed those his words, which he formerly spake, that this happy island where we now stood was known to few, and yet knew most of the nations of the world; which we found to be true, considering they had the languages of Europe, and knew much of our state[89] and business; and yet we in Europe (notwithstanding all the remote discoveries and navigations of this last age,) never heard any of the least inkling or glimpse of this island. This we found wonderful strange; for that all nations have inter-knowledge one of another either by voyage into foreign parts, or by strangers that come to them: and though the traveller into a foreign country doth commonly know more by the eye, than he that stayeth at home can by relation of the

[87]*Unusual*, remarkably fine.
[88]*Be* so bold as to.
[89]*Condition*.

traveller; yet both ways suffice to make a mutual knowledge, in some degree, on both parts. But for this island, we never heard tell of any ship of theirs that had been seen to arrive upon any shore of Europe; no, nor of either the East or West Indies;[90] nor yet of any ship of any other part of the world that had made return from them. And yet the marvel rested not in this. For the situation[91] of it (as his lordship said) in the secret conclave[92] of such a vast sea might cause it. But then that they should have knowledge of the languages, books, affairs, of those that lie such a distance from them, it was a thing we could not tell what to make of; for that it seemed to us a condition and propriety[93] of divine powers and beings, to be hidden and unseen to others, and yet to have others open and as in a light to them." At this speech the governor gave a gracious smile, and said; "That we did well to ask pardon for this question we now asked; for that it imported as if[94] we thought this land a land of magicians, that sent forth spirits of the air into all parts, to bring them news and intelligence of other countries." It was answered by us all, in all possible humbleness, but yet with a countenance taking knowledge[95] that we knew that he spake it but merrily, "That we were apt enough to think there was somewhat supernatural in this island; but yet rather as angelical than magical. But to let his lordship know truly what it was that made us tender[96] and doubtful[97] to ask this question, it was not any such conceit, but because we remembered he had given a touch in his former speech, that this land had laws of secrecy touching strangers." To this he said; "You remember it aright; and therefore in that I shall say to you I must reserve some

[90]*East Indies*: India and adjacent regions and islands; West Indies: lands of Western Hemisphere discovered in the 15th and 16th centuries.
[91]*Location*.
[92]*Private* room, inner chamber, closet.
[93]*Special quality*.
[94]*Implied*, signified that.
[95]*A* gesture showing.
[96]*Cautious*.
[97]*Apprehensive*.

particulars, which it is not lawful for me to reveal; but there will be enough left to give you satisfaction.

"You shall understand (that which perhaps you will scarce think credible) that about three thousand years ago, or somewhat more, the navigation of the world, (specially for remote voyages,) was greater than at this day. Do not think with yourselves that I know not how much it is increased with you within these six-score years:[98] I know it well: and yet I say greater then than now; whether it was, that the example of the ark, that saved the remnant of men from the universal deluge, gave men confidence to adventure upon the waters; or what it was; but such is the truth. The Phœnicians,[99] and especially the Tyrians,[100] had great fleets. So had the Carthaginians,[101] their colony, which is yet further west. Toward the east, the shipping of Egypt and of Palestina[102] was likewise great. China also, and the great Atlantis (that you call America), which have now but junks and canoes, abounded then in tall ships. This island (as appeareth by faithful registers of those times) had then fifteen hundred strong ships, of great content.[103] Of all this there is with you sparing memory, or none; but we have large knowledge thereof.

"At that time, this land was known and frequented by the ships and vessels of all the nations before named. And (as it cometh to pass) they had many times men of other countries, that were no sailors, that came with them; as Persians, Chaldeans,[104] Arabians; so as almost all nations of might and fame

[98]*Taking* the year 1492 as the date of the opening of the New World to European navigation, the date of the sailors' voyage is six-score (120) years later; *i.e.*, 1612.

[99]*Inhabitants* of the ancient country north of Palestine on the Syrian coast.

[100]*Inhabitants* of the ancient Phoenician city of Tyre.

[101]*Phoenician* colonists of the ancient city of North Africa, now Tunis.

[102]*Palestine.*

[103]*Tonnage,* capacity.

[104]*Inhabitants* of ancient Mesopotamia.

resorted hither; of whom we have some stirps[105] and little tribes with us at this day. And for our own ships, they went sundry voyages, as well to your Straits, which you call the Pillars of Hercules, as to other parts in the Atlantic and Mediterrane Seas; as to Paguin (which is the same with Cambaline) and Quinzy, upon the Oriental Seas, as far as to the borders of the East Tartary.[106]

"At the same time, and an age after, or more, the inhabitants of the great Atlantis did flourish. For though the narration and description which is made by a great man[107] with you, that the descendants of Neptune[108] planted[109] there; and of the magnificent temple, palace, city, and hill; and the manifold streams of goodly navigable rivers, (which, as so many chains, environed the same site and temple); and the several degrees[110] of ascent whereby men did climb up to the same, as if it had been a *scala cœli*;[111] be all poetical and fabulous: yet so much is true, that the said country of Atlantis, as well that of Peru, then called Coya, as that of Mexico,[112] then named Tyrambel, were mighty and proud kingdoms in arms, shipping, and riches: so mighty, as at one time (or at least within the space of ten years) they both made two great expeditions; they of Tyrambel through the Atlantic to the Mediterrane Sea; and they of Coya through the South Sea upon this our island. And for the former of these, which was into Europe, the same author amongst you (as it seemeth) had some relation from the

---

[105]*Lineages.*
[106]*Pillars* of Hercules: the Straits of Gibraltar; Paguin, Cambaline: names for Peking; Quinzy: Hangchow; Borders of East Tartary: East Coast of Asia, north of China.
[107]*Plato Timaeus* 21e1–25d7; *Critias* 113a1–121c4.
[108]*God* of the sea in Roman mythology corresponding to the Greek god Poseidon.
[109]*Settled.*
[110]*Steps.*
[111]*Ladder* of heaven.
[112]*The* ancient Atlantis was apparently in North America; Coya and Tyrambel were in Central and South America.

Egyptian priest whom he citeth.[113] For assuredly such a thing there was. But whether it were the ancient Athenians that had the glory of the repulse and resistance of those forces, I can say nothing: but certain it is, there never came back either ship or man from that voyage. Neither had the other voyage of those of Coya upon us had better fortune,[114] if they had not met with enemies of greater clemency. For the king of this island (by name Altabin)[115] a wise man and a great warrior, knowing well both his own strength and that of his enemies, handled the matter so, as he cut off their land-forces from their ships; and entoiled[116] both their navy and their camp with a greater power than theirs, both by sea and land; and compelled them to render themselves without striking stroke: and after they were at his mercy, contenting himself only with their oath that they should no more bear arms against him, dismissed them all in safety. But the Divine Revenge overtook not long after those proud enterprises. For within less than the space of one hundred years, the great Atlantis was utterly lost and destroyed: not by a great earthquake, as your man saith,[117] (for that whole tract is little subject to earthquakes,) but by a particular deluge or inundation;[118] those countries having, at this day, far greater rivers and far higher mountains to pour down waters, than any part of the old world.[119] But it is true that the same inundation was not deep; not past forty foot, in most places, from the ground: so that although it destroyed man and beast generally, yet some few wild inhabitants of the

[113]*Timaeus* 21a7 ff.; *Critias* 110a6 ff.
[114]*The* Coyans would have fared no better.
[115]*Combination* of Latin words meaning "twice lofty."
[116]*Entrapped*.
[117]*Timaeus* 25c6.
[118]*Not* the universal flood survived by Noah.
[119]*In* Plato's account the Athenian warriors were destroyed by earthquakes and floods that caused the whole island of Atlantis to be swallowed by the sea. The war no longer lived in Athenian memory because Athens, along with the rest of the world except for the Egyptians, was subject to various destructions of mankind caused by fire, water, and other means; *Timaeus* 21d2–23d1.

wood escaped. Birds also were saved by flying to the high trees and woods. For as for men, although they had buildings in many places higher than the depth of the water, yet that inundation, though it were shallow, had a long continuance; whereby they of the vale[120] that were not drowned, perished for want of food and other things necessary. So as[121] marvel you not at the thin population of America, nor at the rudeness and ignorance of the people; for you must account your inhabitants of America as a young people; younger a thousand years, at the least, than the rest of the world; for that there was so much time between the universal flood and their particular inundation. For the poor remnant of human seed which remained in their mountains peopled the country again slowly, by little and little; and being simple and savage people, (not like Noah and his sons, which was the chief family of the earth,) they were not able to leave letters, arts, and civility to their posterity; and having likewise in their mountainous habitations been used (in respect[122] of the extreme cold of those regions) to clothe themselves with the skins of tigers, bears, and great hairy goats, that they have in those parts; when after they came down into the valley, and found the intolerable heats which are there, and knew no means of lighter apparel, they were forced to begin the custom of going naked, which continueth at this day. Only they take great pride and delight in the feathers of birds, and this also they took from those their ancestors of the mountains, who were invited unto it by the infinite flights of birds that came up to the high grounds, while the waters stood below. So you see, by this main accident of time, we lost our traffic with the Americans, with whom of all others, in regard[123] they lay nearest to us, we had most commerce. As for the other parts of the world, it is most manifest that in the ages following (whether it were in respect of wars, or by a

---

[120]*Valley.*
[121]*Therefore.*
[122]*Because.*
[123]*Since.*

natural revolution of time,[124]) navigation did every where greatly decay; and specially far voyages (the rather by the use of galleys, and such vessels as could hardly brook[125] the ocean,) were altogether left and omitted.[126] So then, that part of intercourse which could be from other nations to sail to us, you see how it hath long since ceased; except it were by some rare accident, as this of yours. But now of the cessation of that other part of intercourse, which might be by our sailing to other nations, I must yield you some other cause. For I cannot say (if I shall say truly,) but our shipping, for number, strength, mariners, pilots, and all things that appertain to navigation, is as great as ever: and therefore why we should sit at home, I shall now give you an account by itself: and it will draw nearer[127] to give you satisfaction to your principal question.

"There reigned in this island, about nineteen hundred years ago,[128] a King, whose memory of all others we most adore; not superstitiously, but as a divine instrument, though a mortal man; his name was Solamona: and we esteem him as the lawgiver of our nation. This king had a *large heart*,[129] inscrutable for good;[130] and was wholly bent to make his kingdom and people happy. He therefore, taking into consideration how sufficient and substantive[131] this land was to maintain itself without any aid at all of the foreigner; being five thousand six hundred miles in circuit, and of rare fertility of soil in the greatest part thereof; and finding also the shipping of this country might be plentifully set on work, both by fishing and by transportations from port to port, and likewise by sailing unto some small islands that are not far from us, and are

[124]*The* natural course of time.
[125]*Tolerate.*
[126]*As* navigation decayed, distant voyages were given up.
[127]*Come* closer to giving.
[128]*About* 288 B.C.E.
[129]*Said* of the Biblical Solomon, 1 Kings 4: 29–33, in reference to his wisdom.
[130]*Unfathomably* good, see Prov. 25: 3.
[131]*Independent.*

under the crown and laws of this state; and recalling into his memory the happy and flourishing estate[132] wherein this land then was, so as it might be a thousand ways altered to the worse, but scarce any one way to the better; thought nothing wanted to his noble and heroical intentions, but[133] only (as far as human foresight might reach) to give perpetuity to that which was in his time so happily established. Therefore amongst his other fundamental laws of this kingdom, he did ordain the interdicts[134] and prohibitions which we have touching entrance of strangers; which at that time (though it was after the calamity of America) was frequent; doubting[135] novelties, and commixture of manners. It is true, the like law against the admission of strangers without licence is an ancient law in the kingdom of China, and yet continued in use. But there it is a poor thing; and hath made them a curious,[136] ignorant, fearful, foolish nation. But our lawgiver made his law of another temper.[137] For first, he hath preserved all points of humanity, in taking order[138] and making provision for the relief of strangers distressed; whereof you have tasted." At which speech (as reason was) we all rose up, and bowed ourselves. He went on. "That king also, still desiring to join humanity and policy together; and thinking it against humanity to detain strangers here against their wills, and against policy that they should return and discover[139] their knowledge of this estate,[140] he took this course: he did ordain that of the strangers that should be permitted to land, as many (at all times) might depart as would; but as many as would stay should have very good conditions and means to live from the

---

[132]*Condition.*
[133]*His* noble and heroic intentions needed only.
[134]*An* act of forbidding peremptorily.
[135]*Fearing.*
[136]*Anxious*, solicitous.
[137]*Character.*
[138]*Managing.*
[139]*Reveal.*
[140]*Kingdom*, commonwealth.

state. Wherein he saw so far, that now in so many ages since the prohibition, we have memory not of one ship that ever returned; and but of thirteen persons only, at several[141] times, that chose to return in our bottoms. What those few that returned may have reported abroad I know not. But you must think, whatsoever they have said could be taken where they came but for a dream. Now for our travelling from hence[142] into parts abroad, our Lawgiver thought fit altogether to restrain it. So is it not in China. For the Chineses sail where they will or can; which sheweth that their law of keeping out strangers is a law of pusillanimity and fear. But this restraint of ours hath one only exception, which is admirable; preserving the good which cometh by communicating with strangers, and avoiding the hurt; and I will now open it to you. And here I shall seem a little to digress, but you will by and by find it pertinent. Ye shall understand (my dear friends) that amongst the excellent acts of that king, one above all hath the preeminence. It was the erection and institution of an Order or Society which we call *Salomon's House*; the noblest foundation (as we think) that ever was upon the earth; and the lanthorn[143] of this kingdom. It is dedicated to the study of the Works and Creatures of God. Some think it beareth the founder's name a little corrupted, as if it should be Solamona's House. But the records write it as it is spoken. So as I take it to be denominate[144] of the King of the Hebrews, which is famous with you, and no stranger to us. For we have some parts of his works which with you are lost; namely, that Natural History which he wrote, of all plants, from the *cedar of Libanus*[145] to the *moss that groweth out of the wall*, and of all *things that have life and motion*.[146] This maketh me think that our king, finding himself

---

[141]*Different.*
[142]*Here.*
[143]*Light.*
[144]*Named* for.
[145]*Lebanon.*
[146]*See* note 129 above.

to symbolize in many things with[147] that king of the Hebrews (which lived many years before him), honoured him with the title of this foundation. And I am the rather induced to be of this opinion, for that I find in ancient records this Order or Society is sometimes called Salomon's House and sometimes the College of the Six Days' Works; whereby I am satisfied that our excellent king had learned from the Hebrews that God had created the world and all that therein is within six days; and therefore he instituting that House for the finding out of the true nature of all things, (whereby God might have the more glory in the workmanship of them, and men the more fruit in the use of them,) did give it also that second name. But now to come to our present purpose. When the king had forbidden to all his people navigation into any part that was not under his crown, he made nevertheless this ordinance; That every twelve years there should be set forth out of this kingdom two ships, appointed to several[148] voyages; That in either[149] of these ships there should be a mission of three of the Fellows or Brethren of Salomon's House; whose errand was only to give us knowledge of the affairs and state of those countries to which they were designed,[150] and especially of the sciences, arts, manufactures, and inventions of all the world; and withal[151] to bring unto us books, instruments, and patterns in every kind; That the ships, after they had landed the brethren, should return; and that the brethren should stay abroad till the new mission. These ships are not otherwise fraught,[152] than with store of victuals, and good quantity of treasure to remain with the brethren, for the buying of such things and rewarding of such persons as they should think fit. Now for me to tell you how the vulgar sort[153]

---

[147]*To* resemble in many things.
[148]*Different.*
[149]*Each.*
[150]*Bound* for.
[151]*In* addition.
[152]*Laden.*
[153]*Common.*

of mariners are contained[154] from being discovered at land; and how they that must be put on shore for any time, colour[155] themselves under the names of other nations; and to what places these voyages have been designed; and what places of *rendez-vous*[156] are appointed for the new missions; and the like circumstances of the practique;[157] I may not do it: neither is it much to your desire. But thus you see we maintain a trade, not for gold, silver, or jewels; nor for silks; nor for spices; nor any other commodity of matter; but only for God's first creature, which was *Light*: to have *light*[158] (I say) of the growth of all parts of the world." And when he had said this, he was silent; and so were we all. For indeed we were all astonished to hear so strange things so probably[159] told. And he, perceiving that we were willing to say somewhat[160] but had it not ready, in great courtesy took us off,[161] and descended[162] to ask us questions of our voyage and fortunes; and in the end concluded, that we might do well to think with ourselves what time of stay we would demand of the state; and bade us not to scant[163] ourselves; for he would procure such time as we desired. Whereupon we all rose up, and presented ourselves to kiss the skirt of his tippet; but he would not suffer us;[164] and so took his leave. But when it came once amongst our people[165] that the state used to offer conditions[166] to strangers that would stay, we had work enough to get any of our men to look to our

---

[154]*Kept.*
[155]*Misrepresent*, disguise.
[156]*Meeting.*
[157]*Actual* practice.
[158]*Knowledge.*
[159]*Plausibly.*
[160]*Something.*
[161]*Relieved* us.
[162]*Lowered* himself.
[163]*Stint.*
[164]*Allow* us to do it.
[165]*As* soon as our people heard.
[166]*Good* circumstances.

ship, and to keep them from going presently to the governor to crave[167] conditions. But with much ado we refrained them, till we might agree what course to take.

We took ourselves now for free men, seeing there was no danger of our utter perdition; and lived most joyfully, going abroad and seeing what was to be seen in the city and places adjacent within our tedder;[168] and obtaining acquaintance with many of the city, not of the meanest quality;[169] at whose hands we found such humanity, and such a freedom and desire to take strangers as it were into their bosom, as was enough to make us forget all that was dear to us in our own countries: and continually we met with many things right worthy of observation and relation; as indeed, if there be a mirror[170] in the world worthy to hold men's eyes, it is that country. One day there were two of our company bidden to a Feast of the Family, as they call it. A most natural, pious, and reverend custom it is, shewing that nation to be compounded of all goodness. This is the manner of it. It is granted to any man that shall live to see thirty persons descended of his body alive together, and all above three years old, to make this feast; which is done at the cost of the state. The Father of the Family, whom they call the *Tirsan*,[171] two days before the feast, taketh to him three of such friends as he liketh to choose; and is assisted also by the governor of the city or place where the feast is celebrated; and all the persons of the family, of both sexes, are summoned to attend him. These two days the Tirsan sitteth in consultation concerning the good estate[172] of the family. There, if there be any discord or suits beween any of the family, they are compounded and appeased. There, if any of the family

---

[167]*To* beg for.

[168]*Tether.*

[169]*Poorest,* lowest class.

[170]*Model* of excellence.

[171]*Tirsan* is a Persian word (tarsān) meaning timid or fearful.

[172]*Condition.*

be distressed[173] or decayed,[174] order is taken for their relief and competent means to live. There, if any be subject to vice, or take ill courses,[175] they are reproved and censured. So likewise direction is given touching marriages, and the courses of life which any of them should take, with divers other the like orders and advices. The governor assisteth, to the end to put in execution by his public authority the decrees and orders of the Tirsan, if they should be disobeyed; though that seldom needeth; such reverence and obedience they give to the order of nature. The Tirsan doth also then ever choose one man from amongst his sons, to live in house with him: who is called ever after the Son of the Vine. The reason will hereafter appear. On the feast-day, the Father or Tirsan cometh forth after divine service into a large room where the feast is celebrated; which room hath an half-pace[176] at the upper end. Against the wall, in the middle of the half-pace, is a chair placed for him, with a table and carpet before it. Over the chair is a state,[177] made round or oval, and it is of ivy; an ivy somewhat whiter than ours, like the leaf of a silver asp,[178] but more shining; for it is green all winter. And the state is curiously wrought with silver and silk of divers colours, broiding[179] or binding in the ivy; and is ever of the work of some of the daughters of the family; and veiled over at the top with a fine net of silk and silver. But the substance of it is true ivy; whereof, after it is taken down, the friends of the family are desirous to have some leaf or sprig to keep. The Tirsan cometh forth with all his generation or lineage, the males before him, and the females following him; and if there be a mother from whose body the whole lineage is descended, there is a traverse[180] placed in a loft above

---

[173]*In* difficult financial condition.
[174]*Behind* in paying rent.
[175]*Act* unwisely.
[176]*Dais.*
[177]*Canopy.*
[178]*Aspen,* poplar.
[179]*Interweaving.*
[180]*Screened* compartment.

on the right hand of the chair, with a privy[181] door, and a carved window of glass, leaded with gold and blue; where she sitteth, but is not seen. When the Tirsan is come forth, he sitteth down in the chair; and all the lineage place themselves against the wall, both at his back and upon the return[182] of the half-pace, in order of their years without difference of sex; and stand upon their feet. When he is set; the room being always full of company, but well kept and without disorder; after some pause there cometh in from the lower end of the room a *Taratan* (which is as much as an herald) and on either side of him two young lads; whereof one carrieth a scroll of their shining yellow parchment; and the other a cluster of grapes of gold, with a long foot or stalk. The herald and children are clothed with mantles of sea-water green sattin; but the herald's mantle is streamed[183] with gold, and hath a train. Then the herald with three curtesies,[184] or rather inclinations,[185] cometh up as far as the half-pace; and there first taketh into his hand the scroll. This scroll is the King's Charter, containing gift of revenew,[186] and many privileges, exemptions, and points of honour, granted to the Father of the Family; and is ever styled and directed, *To such an one our well-beloved friend and creditor*: which is a title proper only to this case. For they say the king is debtor to no man, but for propagation of his subjects.[187] The seal set to the king's charter is the king's image, imbossed or moulded in gold; and though such charters be expedited of course,[188] and as of right, yet they are varied by discretion, according to the number and dignity of the family. This charter the herald readeth aloud; and while it is read, the father or Tirsan standeth up, supported by two of his sons, such as he

---

[181]*Concealed.*
[182]*Side.*
[183]*Striped.*
[184]*Bows.*
[185]*Bowings* or nods of the head.
[186]*Income.*
[187]*The* only thing the king needs is the propagation of his subjects.
[188]*Officially* issued or granted.

chooseth. Then the herald mounteth the half-pace, and delivereth the charter into his hand: and with that there is an acclamation by all that are present in their language, which is thus much:[189] *Happy are the people of Bensalem.* Then the herald taketh into his hand from the other child the cluster of grapes, which is of gold, both the stalk and the grapes. But the grapes are daintily enamelled; and if the males of the family be the greater number, the grapes are enamelled purple, with a little sun set on the top; if the females, then they are enamelled into a greenish yellow, with a crescent on the top. The grapes are in number as many as there are descendants of the family. This golden cluster the herald delivereth also to the Tirsan; who presently delivereth it over to that son that he had formerly chosen to be in house with him: who beareth it before his father as an ensign[190] of honour when he goeth in public, ever after; and is thereupon called the Son of the Vine. After this ceremony ended, the father or Tirsan retireth; and after some time cometh forth again to dinner, where he sitteth alone under the state, as before; and none of his descendants sit with him, of what degree or dignity soever, except he hap[191] to be of Salomon's House. He is served only by his own children, such as are male; who perform unto him all service of the table upon the knee; and the women only stand about him, leaning against the wall. The room below the half-pace hath tables on the sides for the guests that are bidden; who are served with great and comely order; and towards the end of dinner (which in the greatest feasts with them lasteth never above an hour and an half) there is an hymn sung, varied according to the invention of him that composeth it, (for they have excellent poesy,) but the subject of it is (always) the praises of Adam and Noah and Abraham; whereof the former two peopled the world, and the last was the Father of the Faithful: concluding ever with a thanksgiving for the nativity of our Saviour, in whose birth the births of all are only blessed. Dinner being done, the Tirsan retireth again; and

---

[189]*Which* means roughly.
[190]*Badge.*
[191]*Happen.*

having withdrawn himself alone into a place where he maketh some private prayers, he cometh forth the third time, to give the blessing; with all his descendants, who stand about him as at the first. Then he calleth them forth by one and by one, by name, as he pleaseth, though seldom the order of age be inverted. The person that is called (the table being before removed) kneeleth down before the chair, and the father layeth his hand upon his head, or her head, and giveth the blessing in these words: *Son of Bensalem, (or Daughter of Bensalem,) thy father saith it; the man by whom thou hast breath and life speaketh the word; The blessing of the everlasting Father, the Prince of Peace, and the Holy Dove*[192] *be upon thee, and make the days of thy pilgrimage good and many.*[193] This he saith to every of them; and that done, if there be any of his sons of eminent merit and virtue, (so they be not above two,)[194] he calleth for them again; and saith, laying his arm over their shoulders, they standing; *Sons, it is well ye are born, give God the praise, and persevere to the end.* And withal delivereth to either of them a jewel, made in the figure of an ear of wheat, which they ever after wear in the front of their turban or hat. This done, they fall to music and dances, and other recreations, after their manner, for the rest of the day. This is the full order of that feast.

By that time six or seven days were spent, I was fallen into strait[195] acquaintance with a merchant of that city, whose name was Joabin.[196] He was a Jew, and circumcised: for they have some few stirps of Jews yet remaining among them, whom they leave to their own religion. Which they may the better do, because they are of a far differing disposition from the Jews in

---

[192]*Holy Spirit.*
[193]*Genesis* 47: 9.
[194]*Only* the two best.
[195]*Close.*
[196]*Plural* form of Joab, i.e., "Joabs." Joab was David's nephew and an important captain of David's army. See I Samuel 26: 6; II Samuel 2; 3; 8: 16; 10; 11–14; 17: 25; 18–20; 23–24; I Kings 1–2; 11; I Chronicles 2: 16; 11; 18: 15; 19–21; 26: 28; 27.

other parts. For whereas they hate the name of Christ, and have a secret inbred rancour against the people amongst whom they live: these (contrariwise) give unto our Saviour many high attributes, and love the nation of Bensalem extremely. Surely this man of whom I speak would ever acknowledge that Christ was born of a Virgin, and that he was more than a man; and he would tell how God made him ruler of the Seraphims[197] which guard his throne; and they call him also the *Milken Way*,[198] and the *Eliah* of the *Messiah*;[199] and many other high names; which though they be inferior to his divine Majesty, yet they are far from the language of other Jews. And for the country of Bensalem, this man would make no end of commending it: being desirous, by tradition among the Jews there, to have it believed that the people thereof were of the generations of Abraham, by another son, whom they call Nachoran;[200] and that Moses by a secret cabala[201] ordained the laws of Bensalem which they now use; and that when the Messiah should come, and sit in his throne at Hierusalem,[202] the king of Bensalem should sit at his feet, whereas other kings should keep a great distance. But yet setting aside these Jewish dreams, the man was a wise man, and learned, and of great policy,[203] and excellently seen[204] in the laws and customs of that nation. Amongst other discourses, one day I told him I was much affected[205] with the relation I had from some of the company, of their custom in holding the Feast of the Family; for that (methought) I had never heard of a solemnity wherein nature did so much preside. And because propagation of families

[197]*Six-winged* angels.
[198]*A* way leading to heaven.
[199]*Prophetic* forerunner of the Messiah. See Malachi 4: 5; Matt. 16: 4; 17: 10.
[200]*Genesis* 11: 22–27; 22: 20–23; 24: 15, 47; 29: 5–7; I Chronicles 1: 26; Luke 3: 34.
[201]*Doctrine.*
[202]*Jerusalem.*
[203]*Political* wisdom.
[204]*Well* versed.
[205]*Very* impressed by.

proceedeth from the nuptial copulation, I desired to know of him what laws and customs they had concerning marriage; and whether they kept marriage well; and whether they were tied to one wife? For that where population is so much affected,[206] and such as with them it seemed to be, there is commonly permission of plurality of wives. To this he said, "You have reason for to commend that excellent institution of the Feast of the Family. And indeed we have experience, that those families that are partakers of the blessing of that feast do flourish and prosper ever after in an extraordinary manner. But hear me now, and I will tell you what I know. You shall understand that there is not under the heavens so chaste a nation as this of Bensalem; nor so free from all pollution or foulness. It is the virgin of the world. I remember I have read in one of your European books,[207] of an holy hermit amongst you that desired to see the Spirit of Fornication; and there appeared to him a little foul ugly Æthiop.[208] But if he had desired to see the Spirit of Chastity of Bensalem, it would have appeared to him in the likeness of a fair beautiful Cherubin. For there is nothing amongst mortal men more fair and admirable, than the chaste minds of this people. Know therefore, that with them there are no stews,[209] no dissolute houses,[210] no courtesans, nor any thing of that kind. Nay they wonder (with detestation) at you in Europe, which permit such things. They say ye have put marriage out of office: for marriage is ordained a remedy for unlawful concupiscence; and natural concupiscence see-meth as a spur to marriage. But when men have at hand a remedy more agreeable to their corrupt will, marriage is almost expulsed. And therefore there are with you seen infinite men that marry not, but chuse rather a libertine and impure single life, than to be yoked in marriage; and many that do marry, marry late, when the prime and strength of their years is past.

[206]*Desired.*
[207]*Sintram* by La Motte Fouqué.
[208]*Blackamoor*, Klein Meister in Sintram.
[209]*Bath* houses.
[210]*Brothels.*

And when they do marry, what is marriage to them but a very bargain; wherein is sought alliance, or portion,[211] or reputation, with some desire (almost indifferent) of issue;[212] and not the faithful nuptial union of man and wife, that was first instituted. Neither is it possible that those that have cast away so basely so much of their strength, should greatly esteem children, (being of the same matter,) as chaste men do. So likewise during marriage, is the case much amended,[213] as it ought to be if those things were tolerated only for necessity? No, but they remain still as a very affront to marriage. The haunting of those dissolute places, or resort to courtesans, are no more punished in married men than in bachelors. And the depraved custom of change,[214] and the delight in meretricious embracements, (where sin is turned into art,) maketh marriage a dull thing, and a kind of imposition or tax. They hear you defend these things, as done to avoid greater evils; as advoutries,[215] deflouring of virgins, unnatural lust, and the like. But they say this is a preposterous wisdom; and they call it *Lot's offer*,[216] who to save his guests from abusing, offered his daughters: nay they say farther that there is little gained in this; for that the same vices and appetites do still remain and abound; unlawful lust being like a furnace, that if you stop the flames altogether, it will quench; but if you give it any vent, it will rage. As for masculine love, they have no touch of it; and yet there are not so faithful and inviolate friendships in the world again as are there; and to speak generally, (as I said before,) I have not read of any such chastity in any people as theirs. And their usual saying is, *That whosoever is unchaste cannot reverence himself*; and they say, *That the reverence of a*

---

[211]*Dowry.*
[212]*Only* a faint desire for children.
[213]*Is* the situation corrected.
[214]*Infidelities* (perhaps wife swapping?). See *Francis Bacon: The Oxford Authors*, ed. Brian Vickers (New York: Oxford University Press), p. 796. Hereafter Vickers.
[215]*Adulteries.*
[216]*See* Genesis 19: 1–11.

man's self is, next religion, the chiefest bridle of all vices." And
when he had said this, the good Jew paused a little; whereupon
I, far more willing to hear him speak on than to speak myself,
yet thinking it decent that upon his pause of speech I should not
be altogether silent, said only this; "That I would say to him, as
the widow of Sarepta said to Elias;[217] that he was come to
bring to memory our sins; and that I confess the righteousness
of Bensalem was greater than the righteousness of Europe." At
which speech he bowed his head, and went on in this manner:
"They have also many wise and excellent laws touching
marriage. They allow no polygamy. They have ordained
that none do intermarry or contract, until a month be passed
from their first interview. Marriage without consent of parents
they do not make void, but they mulct[218] it in the inheritors: for
the children of such marriages are not admitted to inherit above
a third part of their parents' inheritance. I have read in a book
of one of your men, of a Feigned Commonwealth, where the
married couple are permitted, before they contract, to see one
another naked.[219] This they dislike; for they think it a scorn to
give a refusal after so familiar knowledge: but because of many
hidden defects in men and women's bodies, they have a more
civil way; for they have near every town a couple of pools,
(which they call *Adam and Eve's pools*,) where it is permitted
to one of the friends of the man, and another of the friends of
the woman, to see them severally[220] bathe naked."[221]

And as we were thus in conference, there came one that
seemed to be a messenger, in a rich huke,[222] that spake with the
Jew: whereupon he turned to me and said; "You will pardon
me, for I am commanded away in haste." The next morning he
came to me again, joyful as it seemed, and said, "There is word
come to the governor of the city, that one of the Fathers of

[217]*See* I Kings 17.
[218]*Punish* by a fine.
[219]*Thomas* More, *Utopia*; "Their Marriage Customs"; Plato, *Laws*
771e.
[220]*Separately.*
[221]*Cf.* II Samuel 11: 2–27.
[222]*Hooded* cape.

Salomon's House will be here this day seven-night:[223] we have
seen none of them this dozen years. His coming is in state;[224]
but the cause of his coming is secret. I will provide you and your
fellows of a good standing to see his entry." I thanked him, and
told him, "I was most glad of the news." The day being come,
he made his entry. He was a man of middle stature and age,
comely of person, and had an aspect as if he pitied men. He was
clothed in a robe of fine black cloth, with wide sleeves and a
cape. His under garment was of excellent white linen down to
the foot, girt with a girdle of the same; and a sindon or tippet of
the same about his neck. He had gloves that were curious,[225]
and set with stone; and shoes of peach-coloured velvet. His
neck was bare to the shoulders. His hat was like a helmet, or
Spanish Montera;[226] and his locks curled below it decently:[227]
they were of colour brown. His beard was cut round, and of the
same colour with his hair, somewhat lighter. He was carried in
a rich chariot without wheels, litter-wise; with two horses at
either end, richly trapped[228] in blue velvet embroidered; and
two footmen on each side in the like attire. The chariot was all
of cedar, gilt, and adorned with crystal; save that the fore-end
had pannels[229] of sapphires, set in borders of gold, and the
hinder-end the like of emeralds of the Peru colour.[230] There
was also a sun of gold, radiant, upon the top, in the midst; and
on the top before, a small cherub of gold, with wings displayed.
The chariot was covered with cloth of gold tissued[231] upon
blue. He had before him fifty attendants, young men all, in
white sattin loose coats to the mid-leg; and stockings of white
silk; and shoes of blue velvet; and hats of blue velvet; with fine
plumes of divers colours, set round like hat-bands. Next before

[223]*A* week from today.
[224]*With* great pomp and solemnity.
[225]*Elaborate*, carefully made.
[226]*A* round cap.
[227]*Becomingly.*
[228]*Adorned.*
[229]*Panels.*
[230]*Bright* green emeralds; see Vickers, 796.
[231]*Interwoven.*

the chariot went two men, bare-headed, in linen garments down to the foot, girt, and shoes of blue velvet; who carried the one a crosier,[232] the other a pastoral staff[233] like a sheep-hook; neither of them of metal, but the crosier of balm-wood,[234] the pastoral staff of cedar. Horsemen he had none, neither before nor behind his chariot: as it seemeth, to avoid all tumult and trouble. Behind his chariot went all the officers and principals of the Companies of the City.[235] He sat alone, upon cushions of a kind of excellent plush, blue; and under his foot curious carpets of silk of divers colours, like the Persian, but far finer. He held up his bare hand as he went, as blessing the people, but in silence. The street was wonderfully well kept: so that there was never any army had their men stand in better battle-array, than the people stood. The windows likewise were not crowded, but every one stood in them as if they had been placed. When the shew was past, the Jew said to me; "I shall not be able to attend you as I would, in regard of some charge the city hath laid upon me, for the entertaining of this great person." Three days after, the Jew came to me again, and said; "Ye are happy men; for the Father of Salomon's House taketh knowledge of your being here, and commanded me to tell you that he will admit all your company to his presence, and have private conference with one of you that ye shall choose: and for this hath appointed the next day after to-morrow. And because he meaneth to give you his blessing, he hath appointed it in the forenoon." We came at our day and hour, and I was chosen by my fellows for the private access. We found him in a fair chamber, richly hanged,[236] and carpeted under foot, without any degrees[237] to the state.[238] He was set

---

[232]*Cross* of an archbishop.
[233]*Staff* borne by a bishop.
[234]*Balsam.*
[235]*Trade* guilds.
[236]*Adorned* with tapestry.
[237]*Steps.*
[238]*Throne.*

upon a low throne richly adorned, and a rich cloth of state[239] over his head, of blue satin embroidered. He was alone, save that he had two pages of honour, on either hand one, finely attired in white. His undergarments were the like that we saw him wear in the chariot; but instead of his gown, he had on him a mantle with a cape, of the same fine black, fastened about him. When we came in, as we were taught, we bowed low at our first entrance; and when we were come near his chair, he stood up, holding forth his hand ungloved, and in posture of blessing; and we every one of us stooped down, and kissed the hem of his tippet. That done, the rest departed, and I remained. Then he warned the pages forth of the room, and caused me to sit down beside him, and spake to me thus in the Spanish tongue:

"God bless thee, my son; I will give thee the greatest jewel I have. For I will impart unto thee, for the love of God and men, a relation of the true state of Salomon's House. Son, to make you know the true state of Salomon's House, I will keep this order. First, I will set forth unto you the end of our foundation. Secondly, the preparations and instruments we have for our works. Thirdly, the several employments and functions whereto our fellows are assigned. And fourthly, the ordinances and rites which we observe.

"The End of our Foundation is the knowledge of Causes, and secret motions of things; and the enlarging of the bounds of Human Empire, to the effecting of all things possible.

"The Preparations and Instruments are these. We have large and deep caves of several depths: the deepest are sunk six hundred fathom; and some of them are digged and made under great hills and mountains: so that if you reckon together the depth of the hill and the depth of the cave, they are (some of them) above three miles deep. For we find that the depth of a hill, and the depth of a cave from the flat, is the same thing;

[239]*Official* cloth.

both remote alike from the sun and heaven's beams, and from the open air.²⁴⁰ These caves we call the Lower Region. And we use them for all coagulations, indurations,²⁴¹ refrigerations, and conservations of bodies. We use them likewise for the imitation of natural mines;²⁴² and the producing also of new artificial metals, by compositions and materials which we use, and lay there for many years. We use them²⁴³ also sometimes, (which may seem strange,) for curing of some diseases, and for prolongation of life in some hermits that choose to live there, well accommodated of²⁴⁴ all things necessary; and indeed live very long; by whom also we learn many things.

"We have burials in several earths, where we put divers cements, as the Chineses do their porcellain. But we have them in greater variety, and some of them more fine. We have also great variety of composts, and soils, for the making of the earth fruitful.

"We have high towers; the highest about half a mile in height; and some of them likewise set upon high mountains; so that the vantage²⁴⁵ of the hill with the tower is in the highest of them three miles at least. And these places we call the Upper Region: accounting the air between the high places and the low, as a Middle Region. We use these towers, according to their several heights and situations, for insolation,²⁴⁶ refrigeration, conservation; and for the view of divers meteors;²⁴⁷ as winds, rain, snow, hail; and some of the fiery meteors²⁴⁸ also. And upon them, in some places, are dwellings of hermits, whom we visit sometimes, and instruct what to observe.

---

²⁴⁰*Thus*, caves dug downward under hills are especially deep.
²⁴¹*Hardenings*.
²⁴²*Mineral* veins.
²⁴³*The* caves.
²⁴⁴*Well* provided with.
²⁴⁵*Total* height.
²⁴⁶*Exposure* to the sun.
²⁴⁷*Atmospheric* phenomena.
²⁴⁸*Shooting* stars and lightning.

"We have great lakes both salt and fresh, whereof we have use for the fish and fowl. We use them also for burials of some natural bodies: for we find a difference in things buried in earth or in air below the earth, and things buried in water. We have also pools, of which some do strain fresh water out of salt; and others by art do turn fresh water into salt. We have also some rocks in the midst of the sea, and some bays upon the shore, for some works wherein is required the air and vapour of the sea. We have likewise violent streams and cataracts,[249] which serve us for many motions: and likewise engines for multiplying and enforcing of winds, to set also on going[250] divers motions.

"We have also a number of artificial wells and fountains, made in imitation of the natural sources and baths; as tincted upon[251] vitriol,[252] sulphur, steel, brass, lead, nitre,[253] and other minerals. And again we have little wells for infusions[254] of many things, where the waters take the virtue[255] quicker and better than in vessels or basons. And amongst them we have a water which we call Water of Paradise, being, by that we do to it, made very sovereign[256] for health, and prolongation of life.

"We have also great and spacious houses, where we imitate and demonstrate meteors; as snow, hail, rain, some artificial rains of bodies and not of water, thunders, lightnings; also generations of bodies in air; as frogs, flies, and divers others.

"We have also certain chambers, which we call Chambers of Health, where we qualify[257] the air as we think good and proper for the cure of divers diseases, and preservation of health.

[249] *Steep* waterfalls.
[250] *To* produce.
[251] *Tinged* with.
[252] *Sulphate* of metal.
[253] *Saltpeter*.
[254] *Modifications* by mixing in new elements.
[255] *Medicinal* efficacy.
[256] *Most* efficacious.
[257] *Alter*, change.

"We have also fair and large baths, of several mixtures, for the cure of diseases, and the restoring of man's body from arefaction:[258] and others for the confirming[259] of it in strength of sinews, vital parts, and the very juice and substance of the body.

"We have also large and various orchards and gardens, wherein we do not so much respect beauty, as variety of ground and soil, proper for divers trees and herbs: and some very spacious, where trees and berries are set whereof we make divers kinds of drinks, besides the vineyards. In these we practise likewise all conclusions[260] of grafting and inoculating,[261] as well of wild-trees as fruit-trees, which produceth many effects. And we make (by art) in the same orchards and gardens, trees and flowers to come[262] earlier or later than their seasons; and to come up and bear more speedily than by their natural course they do. We make them also by art greater much than their nature; and their fruit greater and sweeter and of differing taste, smell, colour, and figure, from their nature. And many of them we so order,[263] as they become of medicinal use.

"We have also means to make divers plants rise by mixtures of earths without seeds; and likewise to make divers new plants, differing from the vulgar;[264] and to make one tree or plant turn into another.

"We have also parks and inclosures of all sorts of beasts and birds, which we use not only for view or rareness,[265] but likewise for dissections and trials; that thereby we may take light[266] what may be wrought upon[267] the body of man. Wherein we find many strange effects; as continuing life in

[258]*Withering.*
[259]*Strengthening.*
[260]*Experiments.*
[261]*Budding.*
[262]*Germinate.*
[263]*Treat* as medicines; see Vickers, 798.
[264]*Common.*
[265]*For* display and preservation of rare species.
[266]*Learn.*
[267]*Done* to.

them, though divers parts, which you account vital, be perished and taken forth; resuscitating of some that seem dead in appearance; and the like. We try also all poisons and other medicines upon them, as well of chirurgery[268] as physic.[269] By art likewise, we make them greater or taller than their kind is; and contrariwise dwarf them, and stay their growth: we make them more fruitful and bearing than their kind is; and contrariwise barren and not generative. Also we make them differ in colour, shape, activity, many ways. We find means to make commixtures and copulations[270] of different kinds; which have produced many new kinds, and them not barren, as the general opinion is. We make a number of kinds of serpents, worms, flies, fishes, of putrefaction,[271] whereof some are advanced (in effect) to be perfect creatures,[272] like beasts or birds; and have sexes, and do propagate. Neither do we this by chance, but we know beforehand of what matter and commixture what kind of those creatures will arise.

"We have also particular pools, where we make trials upon fishes, as we have said before of beasts and birds.

"We have also places for breed and generation of those kinds of worms and flies which are of special use; such as are with you your silk-worms and bees.

"I will not hold you long with recounting of our brew-houses, bake-houses, and kitchens, where are made divers drinks, breads, and meats, rare and of special effects. Wines we have of grapes; and drinks of other juice of fruits, of grains, and of roots: and of mixtures with honey, sugar, manna, and fruits dried and decocted.[273] Also of the tears[274] or woundings of trees, and of the pulp of canes. And these drinks are of

[268]*Surgery.*
[269]*Medicine.*
[270]**Mixings** and couplings.
[271]**Out** of decaying matter.
[272]**Complex** enough to reproduce.
[273]**Condensed.**
[274]**Drops** of sap exuded spontaneously.

several ages, some to the age or last[275] of forty years. We have drinks also brewed with several herbs, and roots, and spices; yea with several fleshes, and white meats; whereof some of the drinks are such, as they are in effect meat and drink both: so that divers, especially in age,[276] do desire to live with them, with little or no meat or bread. And above all, we strive to have drinks of extreme thin parts,[277] to insinuate[278] into the body, and yet without all[279] biting, sharpness, or fretting;[280] insomuch as some of them put upon the back of your hand will, with a little stay,[281] pass through to the palm, and yet taste mild to the mouth. We have also waters which we ripen[282] in that fashion, as they become nourishing; so that they are indeed excellent drink; and many will use no other. Breads we have of several grains, roots, and kernels: yea and some of flesh and fish dried; with divers kinds of leavenings and seasonings: so that some do extremely move appetites;[283] some do nourish so, as divers[284] do live of them, without any other meat; who live very long. So for meats, we have some of them so beaten and made tender and mortified,[285] yet without all corrupting,[286] as a weak heat of the stomach will turn them into good chylus,[287] as well as a strong heat would meat otherwise prepared. We have some meats also and breads and drinks, which taken by men enable them to fast long after; and some other, that used make

[275]*Duration.*
[276]*Various* people, especially the aged.
[277]*Extremely* fine composition.
[278]*Penetrate* subtly.
[279]*Any.*
[280]*Corrosion.*
[281]*If* allowed to remain there a short time.
[282]*Develop*, perfect.
[283]*Are* very delectable.
[284]*Some* people.
[285]*Tenderized* by hanging.
[286]*Any* rotting.
[287]*Chyle*, fluid in the intestines just before absorption.

the very flesh of men's bodies sensibly[288] more hard and tough, and their strength far greater than otherwise it would be.

"We have dispensatories,[289] or shops of medicines. Wherein you may easily think, if we have such variety of plants and living creatures more than you have in Europe, (for we know what you have,) the simples,[290] drugs, and ingredients of medicines, must likewise be in so much the greater variety. We have them likewise of divers ages, and long fermentations. And for their preparations, we have not only all manner of exquisite distillations and separations, and especially by gentle heats and percolations through divers strainers, yea and substances; but also exact forms of composition, whereby they incorporate[291] almost, as they were natural simples.[292]

"We have also divers mechanical arts, which you have not; and stuffs made by them; as papers, linen, silks, tissues; dainty works of feathers of wonderful lustre; excellent dyes, and many others; and shops likewise, as well for such as are not brought into vulgar use amongst us as for those that are. For you must know that of the things before recited, many of them are grown into use throughout the kingdom; but yet if they did flow from our invention, we have of them also for patterns and principals.[293]

"We have also furnaces of great diversities, and that keep great diversity of heats; fierce and quick; strong and constant; soft and mild; blown, quiet; dry, moist; and the like. But above all, we have heats in imitation of the sun's and heavenly bodies' heats, that pass divers inequalities[294] and (as it were) orbs,[295]

---

[288]*Perceptibly*; see Vickers, 798.
[289]*Dispensaries*.
[290]*Unmixed*, basic substance of medicine, herbs.
[291]*Unite* or blend into new substances.
[292]*Medicines* composed of one ingredient.
[293]*If* not produced then preserved as examples or originals; see Vickers, 798–99.
[294]*Changes* in intensity.
[295]*Cycles*; see Vickers, 799.

progresses,[296] and returns,[297] whereby we produce admirable effects. Besides, we have heats of dungs, and of bellies and maws of living creatures, and of their bloods and bodies; and of hays and herbs laid up moist; of lime unquenched;[298] and such like. Instruments also which generate heat only by motion. And farther, places for strong insolations; and again, places under the earth, which by nature or art yield heat. These divers heats we use, as the nature of the operation which we intend requireth.

"We have also perspective-houses,[299] where we make demonstrations of all lights and radiations; and of all colours; and out of things uncoloured and transparent, we can represent unto you all several colours; not in rain-bows, as it is in gems and prisms, but of themselves single. We represent[300] also all multiplications[301] of light, which we carry to great distance, and make so sharp as to discern small points and lines; also all colorations of light: all delusions and deceits[302] of the sight, in figures, magnitudes, motions, colours: all demonstrations of shadows. We find also divers means, yet unknown to you, of producing of light originally from divers bodies. We procure means of seeing objects afar off; as in the heaven and remote places; and represent things near as afar off, and things afar off as near; making feigned distances. We have also helps for the sight, far above spectacles and glasses in use. We have also glasses and means to see small and minute bodies perfectly and distinctly; as the shapes and colours of small flies and worms, grains and flaws in gems, which cannot otherwise be seen; observations in urine and blood, not otherwise to be seen. We make artificial rain-bows, halos, and circles about light. We

---

[296]*Forward* courses.
[297]*Return* courses.
[298]*Unslaked.*
[299]*Houses* for optical instruments.
[300]*Produce.*
[301]*Intensifications.*
[302]*Illusions.*

represent also all manner of reflexions, refractions, and multi-plications of visual beams of objects.

"We have also precious stones of all kinds, many of them of great beauty, and to you unknown; crystals likewise; and glasses of divers kinds; and amongst them some of metals vitrificated,[303] and other materials besides those of which you make glass. Also a number of fossils,[304] and imperfect miner-als, which you have not. Likewise loadstones[305] of prodigious virtue;[306] and other rare stones, both natural and artificial.

"We have also sound-houses, where we practise and dem-onstrate all sounds, and their generation. We have harmonies which you have not, of quarter-sounds,[307] and lesser slides[308] of sounds. Divers instruments of music likewise to you unknown, some sweeter than any you have; together with bells and rings[309] that are dainty and sweet. We represent small sounds as great and deep; likewise great sounds extenuate[310] and sharp; we make divers tremblings and warblings of sounds, which in their original are entire.[311] We represent and imitate all articulate sounds and letters, and the voices and notes of beasts and birds. We have certain helps which set to the ear do further the hearing greatly. We have also divers strange and artificial echos, reflecting the voice many times, and as it were tossing it: and some that give back the voice louder than it came; some shriller, and some deeper; yea, some rendering the voice differing in the letters or articulate sound from that they receive. We have also means to convey sounds in trunks[312] and pipes, in strange lines[313] and distances.

[303]*Converted* into a glass-like substance.
[304]*Any* rock or mineral dug out of the earth.
[305]*Magnets.*
[306]*Power.*
[307]*Quarter-notes.*
[308]*Musical* intervals; see Vickers, 799.
[309]*Chimes.*
[310]*Thin.*
[311]*Whole.*
[312]*Tubes.*
[313]*Irregular* lines; see Vickers, 799.

"We have also perfume-houses; wherewith we join also practices[314] of taste. We multiply smells, which may seem strange. We imitate smells, making all smells to breathe out of other mixtures than those that give them. We make divers imitations of taste likewise, so that they will deceive any man's taste. And in this house we contain also a confiture-house;[315] where we make all sweetmeats,[316] dry and moist, and divers pleasant wines, milks, broths, and sallets,[317] far in greater variety than you have.

"We have also engine-houses, where are prepared engines and instruments for all sorts of motions. There we imitate and practise to make swifter motions than any you have, either out of your muskets or any engine that you have; and to make them and multiply them more easily, and with small force, by wheels and other means: and to make them stronger, and more violent than yours are; exceeding your greatest cannons and basilisks.[318] We represent also ordnance and instruments of war, and engines of all kinds: and likewise new mixtures and compositions of gun-powder, wildfires burning in water, and unquenchable. Also fire-works of all variety both for pleasure and use. We imitate also flights of birds; we have some degrees[319] of flying in the air; we have ships and boats for going under water, and brooking of seas; also swimming-girdles[320] and supporters. We have divers curious[321] clocks, and other like motions of return,[322] and some perpetual motions. We imitate also motions of living creatures, by images[323] of men, beasts, birds, fishes, and serpents. We

[314]*Experiments*; see Vickers, 799.
[315]*Confection* house.
[316]*Candied* fruits, nuts, etc.
[317]*Salads*.
[318]*Large* cannons named for a deadly serpent.
[319]*Some* success.
[320]*Life* preservers.
[321]*Precise*.
[322]*Machines* producing cyclical motion.
[323]*Robots*, automata.

have also a great number of other various motions, strange for equality, fineness, and subtilty.[324]

"We have also a mathematical house, where are represented all instruments, as well of geometry as astronomy, exquisitely made.

"We have also houses of deceits of the senses; where we represent all manner of feats of juggling, false apparitions, impostures, and illusions; and their fallacies.[325] And surely you will easily believe that we that have so many things truly natural which induce admiration, could in a world of particulars[326] deceive the senses, if we would disguise those things and labour to make them seem more miraculous. But we do hate all impostures and lies: insomuch as[327] we have severely forbidden it to all our fellows, under pain of ignominy and fines, that they do not[328] shew any natural work or thing, adorned or swelling;[329] but only pure as it is, and without all affectation of strangeness.

"These are (my son) the riches of Salomon's House.

"For the several employments and offices of our fellows; we have twelve that sail into foreign countries, under the names of other nations, (for our own we conceal;) who bring us the books, and abstracts, and patterns[330] of experiments of all other parts. These we call Merchants of Light.

"We have three that collect the experiments which are in all books. These we call Depredators.[331]

"We have three that collect the experiments of all mechanical arts; and also of liberal sciences;[332] and also of practices which are not brought into arts.[333] These we call Mystery-men.

[324]*Extraordinary* for regularity, sharpness, and complexity.
[325]*How* they are false.
[326]*That* could in many instances; see Vickers, 800.
[327]*So* much that.
[328]*An* acceptable double negative in Bacon's time.
[329]*Exaggerated*.
[330]*Models*; see Vickers, 800.
[331]*Those* who plunder or pillage.
[332]*Liberal* arts.
[333]*Unsystematic* practices.

"We have three that try new experiments, such as themselves think good. These we call Pioners or Miners.

"We have three that draw the experiments of the former four into titles and tables,[334] to give the better light for[335] the drawing of observations and axioms out of them. These we call Compilers.

"We have three that bend themselves,[336] looking into the experiments of their fellows, and cast about[337] how to draw out of them things of use and practice for man's life, and knowledge as well for works as for plain demonstration[338] of causes, means of natural divinations,[339] and the easy and clear discovery of the virtues[340] and parts of bodies. These we call Dowry-men or Benefactors.

"Then after divers meetings and consults[341] of our whole number, to consider of the former labours and collections, we have three that take care, out of them, to direct new experiments, of a higher light,[342] more penetrating into nature than the former. These we call Lamps.

"We have three others that do execute the experiments so directed, and report them. These we call Inoculators.[343]

"Lastly, we have three that raise the former discoveries by experiments into greater observations, axioms, and aphorisms. These we call Interpreters of Nature.

"We have also, as you must think, novices and apprentices, that the succession of the former employed men do not fail; besides a great number of servants and attendants, men and women. And this we do also: we have consultations, which of the inventions and experiences which we have discovered shall be

[334]*Those* who classify and list the experiments of the former.
[335]*The* better to enable.
[336]*Direct* their attention to.
[337]*Consider.*
[338]*Theoretical* demonstration.
[339]*Discovering* and predicting the secrets of nature.
[340]*Characteristics*, powers.
[341]*Consultations.*
[342]*Producing* more general knowledge.
[343]*Men* who bud trees.

published, and which not: and take all an oath of secrecy, for the concealing of those which we think fit to keep secret: though some of those we do reveal sometimes to the state, and some not.

"For our ordinances and rites: we have two very long and fair galleries: in one of these we place patterns and samples of all manner of the more rare and excellent inventions: in the other we place the statua's[344] of all principal inventors. There we have the statua of your Columbus, that discovered the West Indies: also the inventor of ships: your monk[345] that was the inventor of ordnance and of gunpowder: the inventor of music: the inventor of letters: the inventor of printing: the inventor of observations of astronomy: the inventor of works in metal: the inventor of glass: the inventor of silk of the worm: the inventor of wine: the inventor of corn and bread: the inventor of sugars: and all these by more certain[346] tradition than you have. Then have we divers inventors of our own, of excellent works; which since you have not seen, it were too long to make descriptions of them; and besides, in the right understanding of those descriptions you might easily err. For upon every invention of value, we erect a statua to the inventor, and give him a liberal and honourable reward. These statua's are some of brass; some of marble and touch-stone;[347] some of cedar and other special woods gilt and adorned: some of iron; some of silver; some of gold.

"We have certain hymns and services, which we say daily, of laud,[348] and thanks to God for his marvellous works: and forms of prayers, imploring his aid and blessing for the illumination of our labours, and the turning of them into good and holy uses.

"Lastly, we have circuits or visits of divers principal cities of the kingdom; where, as it cometh to pass, we do publish such

[344]*Statues.*
[345]*Roger* Bacon or Berthold Schwarz.
[346]*More* trustworthy.
[347]*Dark* quartz or jasper.
[348]*In* praise of.

new profitable inventions as we think good. And we do also declare natural divinations of diseases, plagues, swarms of hurtful creatures, scarcity, tempests, earthquakes, great inundations, comets, temperature[349] of the year, and divers other things; and we give counsel thereupon what the people shall do for the prevention and remedy of them."

And when he had said this, he stood up; and I, as I had been taught, kneeled down; and he laid his right hand upon my head, and said; "God bless thee, my son, and God bless this relation which I have made. I give thee leave to publish it for the good of other nations; for we here are in God's bosom, a land unknown." And so he left me; having assigned a value of about two thousand ducats,[350] for a bounty[351] to me and my fellows. For they give great largesses[352] where they come upon all occasions.

[THE REST WAS NOT PERFECTED.]

[349]*Climate.*
[350]*Gold* coins.
[351]*Gift*, gratuity.
[352]*Free* gifts.

OF UNITY IN RELIGION

Religion being the chief band of human society, it is a happy thing when itself is well contained within the true band of Unity. The quarrels and divisions about religion were evils unknown to the heathen. The reason was, because the religion of the heathen consisted rather in rites and ceremonies, than in any constant belief. For you may imagine what kind of faith theirs was, when the chief doctors and fathers of their church were the poets. But the true God hath this attribute, that he is a *jealous God*; and therefore his worship and religion will endure no mixture nor partner. We shall therefore speak a few words concerning the Unity of the Church; what are the fruits thereof; what the bounds; and what the means.

The Fruits of Unity (next unto the well-pleasing of God, which is all in all) are two; the one towards those that are without the church, the other towards those that are within. For the former; it is certain, that heresies and schisms are of all others the greatest scandals; yea, more than corruption of manners. For as in the natural body a wound or solution of continuity is worse than a corrupt humour; so in the spiritual. So that nothing doth so much keep men out of the church, and drive men out of the church, as breach of unity. And therefore, whensoever it cometh to that pass, that one saith, *Ecce in deserto*, another saith, *Ecce in penetralibus*;[1] that is, when some men seek Christ in the conventicles of heretics, and others

[1]*Matt.* 24: 26.

New Atlantis and The Great Instauration, Second Edition.
Edited by Jerry Weinberger.
© 2017 John Wiley & Sons, Inc. Published 2017 by John Wiley & Sons, Inc.

in an outward face of a church, that voice had need continually
to sound in men's ears, *Nolite exire,— go not out.*[2] The Doctor
of the Gentiles (the propriety of whose vocation drew him to
have a special care of those without) saith, *If an heathen come
in, and hear you speak with several tongues, will he not say that
you are mad?*[3] And certainly it is little better, when atheists and
profane persons do hear of so many discordant and contrary
opinions in religion; it doth avert them from the church, and
maketh them *to sit down in the chair of the scorners.* It is but a
light thing to be vouched in so serious a matter, but yet it
expresseth well the deformity. There is a master of scoffing,[4]
that in his catalogue of books of a feigned library sets down this
title of a book, *The morris-dance of Heretics.* For indeed every
sect of them hath a diverse posture or cringe by themselves,
which cannot but move derision in worldlings and depraved
politics, who are apt to contemn holy things.

As for the fruit towards those that are within; it is peace;
which containeth infinite blessings. It establisheth faith. It
kindleth charity. The outward peace of the church distilleth
into peace of conscience. And it turneth the labours of writing
and reading of controversies into treatises of mortification and
devotion.

Concerning the bounds of unity; the true placing of them
importeth exceedingly. There appear to be two extremes. For
to certain zelants all speech of pacification is odious. *Is it peace,
Jehu? What hast thou to do with peace? turn thee behind me.*[5]
Peace is not the matter, but following and party. Contrariwise,
certain Laodiceans[6] and lukewarm persons think they may
accommodate points of religion by middle ways, and taking
part of both, and witty reconcilements; as if they would make
an arbitrement between God and man.[7] Both these extremes

---

[2]*Ibid.*
[3]*1* Cor. 14: 23.
[4]*Rabelais.*
[5]*2* Kings 9: 18–19.
[6]*People* lukewarm or indifferent to religion.
[7]*Col.* 2: 1–3; 4: 13, 15–16; Rev. 3: 14–22.

are to be avoided; which will be done, if the league of Christians penned by our Saviour himself were in the two cross clauses thereof soundly and plainly expounded: H*e that is not with us is against us*; and again, *He that is not against us is with us*;[8] that is, if the points fundamental and of substance in religion were truly discerned and distinguished from points not merely of faith, but of opinion, order, or good intention. This is a thing may seem to many a matter trivial, and done already. But if it were done less partially, it would be embraced more generally.

Of this I may give only this advice, according to my small model. Men ought to take heed of rending God's church by two kinds of controversies. The one is, when the matter of the point controverted is too small and light, not worth the heat and strife about it, kindled only by contradiction. For as it is noted by one of the fathers, *Christ's coat indeed had no seam, but the church's vesture was of divers colours*; whereupon he saith, *In veste varietas sit, scissura non sit*,[9][let there be variety in the garment, but let there be no division:] they be two things, Unity and Uniformity. The other is, when the matter of the point controverted is great, but it is driven to an over-great subtilty and obscurity; so that it becometh a thing rather ingenious than substantial. A man that is of judgment and understanding shall sometimes hear ignorant men differ, and know well within himself that those which so differ mean one thing, and yet they themselves would never agree. And if it come so to pass in that distance of judgment which is between man and man, shall we not think that God above, that knows the heart, doth not discern that frail men in some of their contradictions intend the same thing; and accepteth of both? The nature of such controversies is excellently expressed by St. Paul in the warning and precept that he giveth concerning the same, *Devita profanas vocum novitates, et oppositiones falsi nominis scientæ*:[10]

---

[8]*Matt.* 12: 30; Mark 9: 40; Luke 9: 50, 11: 23.
[9]*St.* Augustine or St. Bernard; see Vickers, 720–21 and John 19: 23–24.
[10]*1* Tim. 6: 20.

[Avoid profane novelties of terms, and oppositions of science falsely so called.] Men create oppositions which are not; and put them into new terms so fixed, as whereas the meaning ought to govern the term, the term in effect governeth the meaning. There be also two false peaces or unities: the one, when the peace is grounded but upon an implicit ignorance; for all colours will agree in the dark: the other, when it is pieced up upon a direct admission of contraries in fundamental points. For truth and falsehood, in such things, are like the iron and clay in the toes of Nebuchadnezzar's image;[11] they may cleave, but they will not incorporate.[12]

Concerning the Means of procuring Unity; men must beware, that in the procuring or muniting[13] of religious unity they do not dissolve and deface the laws of charity and of human society. There be two swords amongst Christians, the spiritual and temporal; and both have their due office and place in the maintenance of religion. But we may not take up the third sword, which is Mahomet's sword, or like unto it; that is, to propagate religion by wars or by sanguinary persecutions to force consciences; except it be in cases of overt scandal, blasphemy, or intermixture of practice against the state; much less to nourish seditions; to authorise conspiracies and rebellions; to put the sword into the people's hands; and the like; tending to the subversion of all government, which is the ordinance of God. For this is but to dash the first table against the second; and so to consider men as Christians, as we forget that they are men. Lucretius the poet, when he beheld the act of Agamemnon, that could endure the sacrificing of his own daughter, exclaimed:

Tantum Relligio potuit suadere malorum:[14]

---

[11]*Dan.* 3: 1–12; 4: 18–27.
[12]*Unite*; see Vickers, 721.
[13]*Fortifying.*
[14]*De rerum natura* 1: 101–102.

[to such ill actions Religion could persuade a man.] What would he have said, if he had known of the massacre in France,[15] or the powder treason of England?[16] He would have been seven times more Epicure and atheist than he was. For as the temporal sword is to be drawn with great circumspection in cases of religion; so it is a thing monstrous to put it into the hands of the common people. Let that be left unto the Anabaptists, and other furies. It was great blasphemy when the devil said, *I will ascend and be like the Highest;*[17] but it is greater blasphemy to personate God, and bring him in saying, *I will descend, and be like the prince of darkness*: and what is it better, to make the cause of religion to descend to the cruel and execrable actions of murthering[18] princes, butchery of people, and subversion of states and governments? Surely this is to bring down the Holy Ghost, instead of the likeness of a dove, in the shape of a vulture or raven; and set out of the bark of a Christian church a flag of a bark of pirates and Assassins. Therefore it is most necessary that the church by doctrine and decree, princes by their sword, and all learnings, both Christian and moral, as by their Mercury rod,[19] do damn and send to hell for ever those facts and opinions tending to the support of the same; as hath been already in good part done. Surely in councils concerning religion, that counsel of the apostle would be prefixed, *Ira hominis non implet justitiam Dei:*[20] [The wrath of man worketh not the righteousness of God.] And it was a notable observation of a wise father,[21] and no less ingenuously confessed; *that those which held and persuaded pressure of*

---

[15]*1572* Catholic massacre of French Protestants on St. Bartholomew's Day
[16]*1605* Catholic plot to blow up Parliament and kill James I.
[17]*Is.* 14: 12–21.
[18]*Murdering*.
[19]*The* rod the Roman god Mercury held when escorting the dead to the underworld.
[20]*Jas.* 1: 19–20.
[21]*See* Vickers, 722.

consciences, were commonly interested[22] *therein themselves for their own ends.*

Synopsis

Bacon's main point in "Of Unity in Religion" is not just that unity is good for religion insofar as religion is the "chief band" of human society in general. While he presents that generalization as true enough, his principal point is that religious unity is especially important for Christianity. The heathen gods, by which Bacon means those of the Greeks, were figments of poets and their worship was a matter of outward "rites and ceremonies." This made Greek worship relatively easy going: the Greeks called on their gods to help them in war (hence Bacon's example of Agamemnon's sacrifice of his daughter), but they did not, Bacon implies, go to war for the sake of religion and religious *doctrine*. Christianity, by contrast, is the religion of a jealous God and not simply a matter of outward ceremonies. Exactly what one believes about God and salvation, doctrine, thus matters very much in Christianity; and that doctrine, being the commands of a jealous god, brooks no others. This fact does not prevent doctrinal disagreements, but it does make such disagreements as inevitably occur prone to extremism and violence. Bacon discusses the "fruits of unity" in two respects: as regards those "without" the church and those "within" it. Nothing makes scoffers and atheists more hostile to the church than the sight of disunity and doctrinal disagreement. And unity gives those within the church peace, faith, and charity. He discusses the two extremes regarding the bonds of religious unity: rigid absolutism and the witty compromising of certain Laodiceans and recommends a middle path between them. Likewise, he discusses two kinds of disagreements especially to be avoided: one is disagreement about petty matters driven only by the spirit of contradiction. The other is disagreement about weighty matters that become

[22] *Interested.*

overly subtle and obscure. In the latter men come to disagree about matters on which they really agree, or create oppositions that don't exist and quarrel over terms that are unclear. Bacon warns, however, against pressing for unity by ignoring the "laws of charity and society," and he especially warns against using wars to propagate religion and "sanguinary persecutions" to "force consciences" or justify sedition, conspiracies, and rebellions, and the "murthering of princes." Bacon concludes by making it clear whereof he speaks: Agamemnon's sacrifice revolted the poet Lucretius, but had that poet, says Bacon, seen the horrors of sectarian conflict in the Christian world he would have been "seven times more Epicure and atheist than he was." Those who would propagate such horrors should be constrained and condemned to hell by princes with their swords and by the church with "doctrine and decree" and "all learnings both Christian and moral."

Questions

Why, in Bacon's view, is Christianity so prone to sectarian conflict?

What do you think of Bacon's conclusion: that princes, the church, and all learning should use force and doctrine to send to hell all those who would use force and persecution to propagate religion and compel consciences?

What Christian religious atrocities does Bacon mention?

OF THE TRUE GREATNESS OF KINGDOMS AND ESTATES

The speech of Themistocles the Athenian, which was haughty and arrogant in taking so much to himself, had been a grave and wise observation and censure, applied at large to others. Desired at a feast to touch a lute, he said, *He could not fiddle, but yet he could make a small town a great city.*[1] These words (holpen[2] a little with a metaphor) may express two differing abilities in those that deal in business of estate. For if a true survey be taken of counsellors and states-men, there may be found (though rarely) those which can make a small state great, and yet cannot fiddle: as, on the other side, there will be found a great many that can fiddle very cunningly, but yet are so far from being able to make a small state great, as their gift lieth the other way; to bring a great and flourishing estate to ruin and decay. And, certainly those degenerate arts and shifts, whereby many counsellors and governors gain both favour with their masters and estimation with the vulgar, deserve no better name than fiddling; being things rather pleasing for the time, and graceful to themselves only, than

---

[1] *Plutarch*, *Themistocles* 2.1–4. See *Cimon* 9.1.
[2] *Helped.*

New Atlantis and The Great Instauration, Second Edition.
Edited by Jerry Weinberger.
© 2017 John Wiley & Sons, Inc. Published 2017 by John Wiley & Sons, Inc.

tending to the weal and advancement of the state which they serve. There are also (no doubt) counsellors and governors which may be held sufficient (*negotiis pares*),[3] able to manage affairs, and to keep them from precipices and manifest inconveniences; which nevertheless are far from the ability to raise and amplify an estate in power, means, and fortune. But be the workmen what they may be, let us speak of the work; that is, the true Greatness of Kingdoms and Estates, and the means thereof. An argument fit for great and mighty princes to have in their hand; to the end that neither by over-measuring their forces, they leese themselves in vain enterprises; nor on the other side, by undervaluing them, they descend to fearful and pusillanimous counsels.

The greatness of an estate in bulk and territory, doth fall under measure; and the greatness of finances and revenue doth fall under computation. The population may appear by musters; and the number and greatness of cities and towns by cards and maps. But yet there is not any thing amongst civil affairs more subject to error, than the right valuation and true judgment concerning the power and forces of an estate. The kingdom of heaven is compared, not to any great kernel or nut, but to a grain of mustard-seed; which is one of the least grains, but hath in it a property and spirit hastily to get up and spread. So are there states great in territory, and yet not apt to enlarge or command; and some that have but a small dimension of stem, and yet apt to be the foundations of great monarchies.

Walled towns, stored arsenals and armories, goodly races of horse, chariots of war, elephants, ordnance, artillery, and the like; all this is but a sheep in a lion's skin, except the breed and disposition of the people be stout and warlike. Nay, number (itself) in armies importeth not much, where the people is of weak courage; for (as Virgil saith) *It never troubles a wolf how many the sheep be*.[4] The army of the Persians in the plains of Arbela was such a vast sea of people, as it did somewhat astonish the commanders in Alexander's army; who came to

[3] *Tacitus*, Annals 6.39, 16.18.
[4] *Virgil*, Eclogues 7.52.

him therefore, and wished him to set upon them by night; but he answered, *He would not pilfer the victory.*[5] And the defeat was easy. When Tigranes, the Armenian, being encamped upon a hill with four hundred thousand men, discovered the army of the Romans, being not above fourteen thousand, marching towards him, he made himself merry with it, and said, *Yonder men are too many for an ambassage, and too few for a fight.*[6] But, before the sun set, he found them enow[7] to give him the chase with infinite slaughter. Many are the examples of the great odds between number and courage: so that a man may truly make a judgment, that the principal point of greatness in any state is to have a race of military men. Neither is money the sinews of war (as it is trivially said), where the sinews of men's arms in base and effeminate people are failing. For Solon said well to Crœsus (when in ostentation he shewed him his gold), *Sir, if any other come that hath better iron than you, he will be master of all this gold.*[8] Therefore let any prince or state think soberly of his forces, except his militia of natives be of good and valiant soldiers. And let princes, on the other side, that have subjects of martial disposition, know their own strength; unless they be otherwise wanting unto themselves. As for mercenary forces (which is the help in this case), all examples show that whatsoever estate or prince doth rest upon them, *he may spread his feathers for a time, but he will mew[9] them soon after.*

The blessing of Judah and Issachar will never meet; *that the same people or nation should be both the lion's whelp and the ass between burthens;*[10] neither will it be, that a people overlaid with taxes should ever become valiant and martial. It is true that taxes levied by consent of the estate do abate men's courage less: as it hath been seen notably in the excises of

[5]*Plutarch, Alexander* 31.12.
[6]*Plutarch, Lucullus* 27.
[7]*Enough.*
[8]*Lucian, Charon* 12.
[9]*Molt.*
[10]*Gen.* 49: 8–14.

the Low Countries; and, in some degree, in the subsidies of England. For you must note that we speak now of the heart and not of the purse. So that although the same tribute and tax, laid by consent or by imposing, be all one to the purse, yet it works diversely upon the courage. So that you may conclude, *that no people over-charged with tribute is fit for empire.*

Let states that aim at greatness, take heed how their nobility and gentlemen do multiply too fast. For that maketh the common subject grow to be a peasant and base swain, driven out of heart, and in effect but the gentleman's labourer. Even as you may see in coppice woods; if you leave your staddles too thick, you shall never have clean underwood, but shrubs and bushes. So in countries, if the gentlemen be too many, the commons will be base; and you will bring it to that, that not the hundred poll will be fit for an helmet; especially as to the infantry, which is the nerve of an army; and so there will be great population and little strength. This which I speak of hath been no where better seen than by comparing of England and France; whereof England, though far less in territory and population, hath been (nevertheless) an overmatch; in regard the middle people of England make good soldiers, which the peasants of France do not. And herein the device of King Henry the Seventh (whereof I have spoken largely in the history of his life) was profound and admirable; in making farms and houses of husbandry of a standard; that is, maintained with such a proportion of land unto them, as may breed a subject to live in convenient plenty and no servile condition; and to keep the plough in the hands of the owners, and not mere hirelings.[11] And thus indeed you shall attain to Virgil's character which he gives to ancient Italy:

Terra potens armis atque ubere glebæ:[12]

---

[11] *See Francis Bacon*: The History of the Reign of King Henry the Seventh: *A New Edition With Introduction, Annotation and Interpretive Essay*, ed. Jerry Weinberger (Ithaca: Cornell University Press, 1996), pp. 84–86, 214–22, 231, 241–44.

[12] *Virgil*, Aeneid 1. 531.

[A land powerful in arms and in productiveness of soil.] Neither is that state (which, for any thing I know, is almost peculiar to England, and hardly to be found any where else, except it be perhaps in Poland) to be passed over; I mean the state of free servants and attendants upon noblemen and gentlemen; which are no ways inferior unto the yeomanry for arms. And therefore out of all question, the splendour and magnificence and great retinues and hospitality of noblemen and gentlemen, received into custom, doth much conduce unto martial greatness. Whereas, contrariwise, the close and reserved living of noblemen and gentlemen causeth a penury of military forces.

By all means it is to be procured, that the trunk of Nebuchadnezzar's tree[13] of monarchy be great enough to bear the branches and the boughs; that is, that the natural subjects of the crown or state bear a sufficient proportion to the stranger subjects that they govern. Therefore all states that are liberal of naturalisation towards strangers are fit for empire. For to think that an handful of people can, with the greatest courage and policy in the world, embrace too large extent of dominion, it may hold for a time, but it will fail suddenly. The Spartans were a nice people in point of naturalisation; whereby, while they kept their compass, they stood firm; but when they did spread, and their boughs were becomen too great for their stem, they became a windfall upon the sudden. Never any state was in this point so open to receive strangers into their body as were the Romans. Therefore it sorted with them accordingly; for they grew to the greatest monarchy. Their manner was to grant naturalisation (which they called *jus civitatis*),[14] and to grant it in the highest degree; that is, not only *jus commercii, jus connubii, jus hæreditatis*; but also *jus suffragii*, and *jus honorum*.[15] And this not to singular persons alone, but likewise to whole families; yea, to cities, and sometimes to nations. Add to

---

[13]*Dan.* 4: 10–26.

[14]*Right* to citizenship.

[15]*Right* to commerce, right to marry, right to inherit property, right to vote, right to hold public office.

this their custom of plantation of colonies; whereby the Roman plant was removed into the soil of other nations. And putting both constitutions together, you will say that it was not the Romans that spread upon the world, but it was the world that spread upon the Romans; and that was the sure way of greatness. I have marvelled sometimes at Spain, how they clasp and contain so large dominions with so few natural Spaniards; but sure the whole compass of Spain is a very great body of a tree; far above Rome and Sparta at the first. And besides, though they have not had that usage to naturalise liberally, yet they have that which is next to it; that is, to employ almost indifferently all nations in their militia of ordinary soldiers; yea and sometimes in their highest commands. Nay it seemeth at this instant they are sensible of this want of natives; as by the Pragmatical Sanction,[16] now published, appeareth.

It is certain, that sedentary and within-door arts, and delicate manufactures (that require rather the finger than the arm), have in their nature a contrariety to a military disposition. And generally, all warlike people are a little idle, and love danger better than travail. Neither must they be too much broken of it, if they shall be preserved in vigour. Therefore it was great advantage in the ancient states of Sparta, Athens, Rome, and others, that they had the use of slaves, which commonly did rid those manufactures. But that is abolished, in greatest part, by the Christian law. That which cometh nearest to it, is to leave those arts chiefly to strangers (which for that purpose are the more easily to be received), and to contain the principal bulk of the vulgar natives within those three kinds,—tillers of the ground; free servants; and handicraftsmen of strong and manly arts, as smiths, masons, carpenters, &c.: not reckoning professed soldiers.

But above all, for empire and greatness, it importeth most, that a nation do profess arms as their principal honour, study, and occupation. For the things which we formerly have spoken of are but habilitations towards arms; and what is habilitation

[16]*Law* proclaimed by King Philip IV of Spain in 1622 to encourage marriage and procreation.

without intention and act? Romulus, after his death (as they report or feign), sent a present to the Romans, that above all they should intend arms; and then they should prove the greatest empire of the world.[17] The fabric of the state of Sparta was wholly (though not wisely) framed and composed to that scope and end. The Persians and Macedonians had it for a flash. The Gauls, Germans, Goths, Saxons, Normans, and others, had it for a time. The Turks have it at this day, though in great declination. Of Christian Europe, they that have it are, in effect, only the Spaniards. But it is so plain *that every man profiteth in that he most intendeth*, that it needeth not to be stood upon. It is enough to point at it; that no nation which doth not directly profess arms, may look to have greatness fall into their mouths. And, on the other side, it is a most certain oracle of time, that those states that continue long in that profession (as the Romans and Turks principally have done) do wonders. And those that have professed arms but for an age, have notwithstanding commonly attained that greatness in that age which maintained them long after, when their profession and exercise of arms hath grown to decay.

Incident to this point is, for a state to have those laws or customs which may reach forth unto them just occasions (as may be pretended) of war. For there is that justice imprinted in the nature of men, that they enter not upon wars (whereof so many calamities do ensue) but upon some, at the least specious, grounds and quarrels. The Turk hath at hand, for cause of war, the propagation of his law or sect; a quarrel that he may always command. The Romans, though they esteemed the extending the limits of their empire to be great honour to their generals when it was done, yet they never rested upon that alone to begin a war. First therefore, let nations that pretend to greatness have this; that they be sensible of wrongs, either upon borderers, merchants, or politic ministers; and that they sit not too long upon a provocation. Secondly, let them be prest and ready to give aids and succours to their confederates; as it ever was with the Romans; insomuch, as if the confederates had

[17] *Plutarch*, *Romulus* 28.

leagues defensive with divers other states, and, upon invasion offered, did implore their aids severally, yet the Romans would ever be the foremost, and leave it to none other to have the honour. As for the wars which were anciently made on the behalf of a kind of party, or tacit conformity of estate, I do not see how they may be well justified: as when the Romans made a war for the liberty of Græcia; or when the Lacedæmonians and Athenians made wars to set up or pull down democracies and oligarchies; or when wars were made by foreigners, under the pretence of justice or protection, to deliver the subjects of others from tyranny and oppression; and the like. Let it suffice, that no estate expect to be great, that is not awake upon any just occasion of arming.

No body can be healthful without exercise, neither natural body nor politic; and certainly to a kingdom, or estate, a just and honourable war is the true exercise. A civil war indeed is like the heat of a fever; but a foreign war is like the heat of exercise, and serveth to keep the body in health; for in a slothful peace, both courages will effeminate and manners corrupt. But howsoever it be for happiness, without all question, for great-ness it maketh, to be still for the most part in arms; and the strength of a veteran army (though it be a chargeable business) always on foot, is that which commonly giveth the law, or at least the reputation, amongst all neighbour states; as may well be seen in Spain, which hath had, in one part or other, a veteran army almost continually, now by the space of six-score years.

To be master of the sea is an abridgement of a monarchy. Cicero, writing to Atticus of Pompey his preparation against Cæsar, saith, *Consilium Pompeii plane Themistocleum est; putat enim, qui mari potitur, eum rerum potiri;* [Pompey is going upon the policy of Themistocles; thinking that he who commands the sea commands all].[18] And, without doubt, Pompey had tired out Cæsar, if upon vain confidence he had not left that way. We see the great effects of battles by sea. The

---

[18]*Cicero, Letters to Atticus,* X. 8 (199.4).

battle of Actium decided the empire of the world.[19] The battle of Lepanto arrested the greatness of the Turk.[20] There be many examples where sea-fights have been final to the war; but this is when princes or states have set up their rest upon the battles. But thus much is certain, that he that commands the sea is at great liberty, and may take as much and as little of the war as he will. Whereas those that be strongest by land are many times nevertheless in great straits. Surely, at this day, with us of Europe, the vantage of strength at sea (which is one of the principal dowries of this kingdom of Great Britain) is great; both because most of the kingdoms of Europe are not merely inland, but girt with the sea most part of their compass; and because the wealth of both Indies seems in great part but an accessory to the command of the seas.

The wars of latter ages seem to be made in the dark, in respect of the glory and honour which reflected upon men from the wars in ancient time. There be now, for martial encouragement, some degrees and orders of chivalry; which nevertheless are conferred promiscuously upon soldiers and no soldiers; and some remembrance perhaps upon the scutcheon; and some hospitals for maimed soldiers; and such like things. But in ancient times, the trophies erected upon the place of the victory; the funeral laudatives and monuments for those that died in the wars; the crowns and garlands personal; the style of Emperor, which the great kings of the world after borrowed; the triumphs of the generals upon their return; the great donatives and largesses upon the disbanding of the armies; were things able to inflame all men's courages. But above all, that of the Triumph, amongst the Romans, was not pageants or gaudery, but one of the wisest and noblest institutions that ever was. For it contained three things; honour to the general; riches to the treasury out of the spoils; and donatives to the army. But that honour perhaps were not fit for monarchies; except it be in the

---

[19]*Naval* battle fought in the Ionian Sea between the forces of Octavian and those of Antony and Cleopatra in 31 B.C.E.
[20]*Naval* battle fought in the Mediterranean Sea between the forces of the Ottoman Empire and those of the Holy League in 1571.

person of the monarch himself, or his sons; as it came to pass in the times of the Roman emperors, who did impropriate the actual triumphs to themselves and their sons, for such wars as they did achieve in person; and left only, for wars achieved by subjects, some triumphal garments and ensigns to the general.

To conclude: no man can *by care taking* (as the Scripture saith) *add a cubit to his stature*,[21] in this little model of a man's body; but in the great frame of kingdoms and commonwealths, it is in the power of princes or estates to add amplitude and greatness to their kingdoms; for by introducing such ordinances, constitutions, and customs, as we have now touched, they may sow greatness to their posterity and succession. But these things are commonly not observed, but left to take their chance.

Synopsis

"Of the True Greatness of Kingdoms and Estates" is Bacon's longest essay. That means, as but a quick glance reveals, that Bacon's longest essay is principally about war. Although Bacon starts out first by distinguishing between merely average counselors and statesmen and those who can make a small state great, and by discussing how to measure the "power and forces of a state" (by which he means the health of its finances and the size of its cities and population), from the third paragraph to the end of the essay he discusses matters of military virtue and war. The means of war possessed by a state may be many and large, but these will prove illusory if "the people is of weak courage." Bacon refers to Alexander, whose smaller forces defeated the Persians and to the Romans, who defeated Tigranes the Armenian's army of four hundred thousand men with a force of just fourteen thousand. Bacon refers to the common argument that "money is the sinews of war" not so much to disagree with it as to point out that money is not sufficient if the sinews of men's arms are failing, as with "base and effeminate" people. A

---

[21]*Matt.* 6: 27; Luke 12: 25.

prince has to rely on his own good soldiers, and so should know well of their condition, and those who depend on mercenaries will have but temporary success. The question, then, is how to produce such good soldiers. The first rule of thumb is not to overtax the people and the second is to be sure that the nobility and gentlemen of a kingdom or estate not be too numerous, lest the people be base and unfit to be soldiers. Bacon's illustration of the second of these rules is the case of England and France, where the smaller population of England, made up predominantly by "middle people," was more than a match for the far more numerous French peasants. This was the result of the policy of King Henry VII, who settled property in such a way as to provide for an independent yeomanry as the source of a powerful infantry. And as regards the smaller English nobility, it was not, as most nobility is, given to "close and reserved living." Rather, the English nobles were accompanied by "free servants" who served as military retainers. Thus did the smaller English nobility also contribute to military greatness. Another, less military means for a state to become great is for it to observe a proper proportion between its citizenry and the people they come to govern, so that states that encourage naturalization are well suited to empire. The Romans were so much this way that the world spread over Rome more than vice versa. And Spain is so aware of this issue of proportion that, in order to govern its empire, it encouraged by law marriage and propagation. And finally, states desiring empire should encourage the manly arts and discourage the practice of dainty and indoor crafts. The best arrangement is for the "vulgar natives" to be farmers, free servants, or crafts-men such as smiths, masons, and carpenters. But above all, says Bacon, a nation aiming at greatness needs to "profess arms" as its principal honor. This was true for Rome and Sparta and the Spaniards and the Turks in our time. To this end states should have laws or policies that give them frequent opportunities for what they can claim are just wars. However, Bacon disap-proves of the wars the ancients fought in order to establish a particular form of government in another state or to liberate the people of another state from what is claimed to be a tyranny. While for a great state war is like the heat of healthy exercise, a civil war is like the fever of a sickness. At any rate, those who have an active standing army, however expensive it might be, will give the law to others. Bacon concludes the essay first by praising states that can fight both on land and sea, since great sea fights settled the Roman Empire and "arrested the greatness of the Turk." He who commands the sea always has greater liberty to take as little or as much of a war as they like. He concludes secondly by contrasting the greater honors and glories the ancients bestowed on great soldiers and generals to the lesser and more promiscuously bestowed honors of the wars of "latter ages." As the Bible says, Bacon concludes, a man cannot add a single "cubit" to his natural height, but with the good laws and policies he has just discussed, a prince or estate can add greatness to their kingdoms and pass it on "to their posterity and succession."

Questions

Why are the English better soldiers than the more numerous French?

Why is money not by itself the means to military power and victory?

What are the non-military elements of the greatness of kingdoms and estates?

On Bacon's New Atlantis

The title and Rawley's opening remarks declaring the incompleteness of the tale are the most obtrusive parts of *New Atlantis*. It is far from obvious why "New Atlantis" should even be the title of the work. "Bensalem," "Renfusa," or "Salomon's House" would all have been more plausible titles, because Atlantis is mentioned only as the presumably kindred neighbor (later destroyed by natural catastrophe) of the country defeated by the ancient Bensalemites. The title thus calls our attention to another important account of Atlantis, to which Bacon refers without actually naming the author. That author is, of course, Plato, who tells the story of Atlantis in his dialogues *Timaeus* and *Critias*.[1] Like *New Atlantis*, Plato's story is incomplete: the dialogue breaks off just as Zeus, angered by the prideful insolence of the technically proficient Atlantis, prepares to announce his intended chastisement of Atlantis to the assembly of the gods. We do not hear about the moderating that was to produce more harmony in the Atlantians.[2] In apparent confirmation of Rawley's opinion, Bacon's story is like Plato's not only in its subject, that of a technically advanced island, but also is similar in its incompleteness. Like Plato's story, Bacon's seems to be missing a speech about politics, a speech that would likely have been similar to Zeus's in considering the themes of punishment and harmony, pride and moderation.

---

[1]*Timaeus* 19b–d, 22a, 25a–b6; *Critias* 120d6–121C4; cf. 106a–b8. See *Phaedrus* 278d7; *Laws* 676ff.
[2]*Critias* 120d6–121c4.

New Atlantis and The Great Instauration, Second Edition.
Edited by Jerry Weinberger.
© 2017 John Wiley & Sons, Inc. Published 2017 by John Wiley & Sons, Inc.

And yet in another respect *New Atlantis* is indeed finished by comparison to Plato's *Critias*. With Plato, the speech of Zeus describing the defects and future perfection of Atlantis is omitted. But with Bacon, the speech of the Father of Salomon's House describing its "true state" as the "noblest foundation" ever upon the earth and the "lanthorn" of Bensalem, is provided, in order to reveal the perfection of *New Atlantis*, "for the good of other nations."[3] Bacon replaces a missing speech about the moderation of excess with a speech about the structure of Salomon's House, the engine of the scientific conquest of nature, and this scientific speech makes no mention of moderation and excess. Yet the Father concludes his speech (and *New Atlantis*) by giving the narrator a massive tip or "largess," despite the fact that throughout the story the Bensalemite's scorn of tipping has been a mark of their apparent moderation and restraint.[4]

The difference between Plato's and Bacon's stories is that Bacon's Bensalem, unlike Plato's Atlantis, has perfected the science that protects her from external harm and nature's corruptibility, indeed from any divine revenge intended by the likes of Zeus. But the two stories are alike in the omission of speeches about excess and moderation. Are we to assume that Bensalemite science by itself assures that the Bensalemites will be harmonious and moderate? Or does the final deed of the story suggest that, despite appearances to the contrary, Bensalem is neither harmonious nor moderate because it lacks the knowledge to be gained from hearkening to Zeus? Perhaps, then, Rawley was both mistaken and yet correct about the completeness of *New Atlantis*: it may be complete because it contains a secret teaching about scientific politics, but it may also be incomplete because that teaching provides no effective practical wisdom about harmony and punishment, or about pride and moderation.

There are two general themes in *New Atlantis*. The first theme presents in two parts what happens to the sailors who

[3]*Above*, pp. 98, 84, 111.
[4]*Above*, pp. 65, 67, 69, 72; cf. 85, 88, 89, 111.

visit Bensalem: we see their plight while they think themselves subject to necessity and chance, and we see their experiences after they come to regard themselves as free men. The second general theme presents in four parts the revelation of Bensalem to the sailors: we see their reception by the Bensalemites, we hear a history of Bensalem, we hear about Bensalemite erotic behavior, and we hear a description of the "House of Salomon."[5] The two general themes and their parts overlap; how they fit together shows what the promise of Bensalem means for the sailors' liberation.

At first glance it seems that in happening upon Bensalem the sailors are freed immediately from dangers beyond their control. But this first impression fades with closer inspection. First, it is possible that the Bensalemites, who have control over wind and sea, have themselves forced the sailors to their shores for purposes of their own. Second, after their initial confinement the sailors are informed by the Bensalemites that they—the sailors—may be told some things they "will not be unwilling to hear," which of course suggests other things the sailors would not have been willing or happy to hear.[6] And finally, the narrator suggests that their initial confinement may be an ominous surveillance and not the harmless custom the Bensalemite official says it is. The narrator's suspicion is confirmed by facts later revealed: According to the priest who is the governor of the Strangers' House, the sailors may anticipate a long and enjoyable stay because the House is well-stocked, it having been thirty-seven years since any strangers had visited the island. In the same vein the priest later remarks—while explaining the Bensalemites' "laws of secrecy" restricting travel in and out of the island—that no strangers have been detained against their will and that the sailors must think that anything reported abroad by returning strangers would be

[5]*The* two sections of the first major theme are: 1) above, pp. 63–87; 2) above, pp. 87–111. The four sections of the second major theme are: 1) above, pp. 63–72; 2) above, pp. 72–87; 3) above, pp. 87–95; 4) above, pp. 95–111.

[6]*Above,* pp. 71–72.

"taken where they came but for a dream." But the lawgiver, Solamona, who promulgated the laws of secrecy, ordained the kindly treatment of strangers permitted on the island precisely because it was "against policy" that strangers "should return and discover [reveal] their knowledge" of Bensalem.[7] That is, the Bensalemite laws concerning strangers and travel, intended to protect the island from foreign corruption, presume the credulousness of non-Bensalemite peoples, contrary to the priest's soothing remarks. The narrator's fears were not unwarranted. If Bensalemite law and policy are consistent, strangers unwilling to stay, or those judged unfit to stay, must have been restrained by force or killed. This doubtless the sailors would not have been willing to hear.

The Bensalemites' initial hospitality and humanity are at the very least ambiguous. And since the sailors represent those who will bring Bensalemite science to the rest of humanity, Bacon makes us consider that the means to scientific happiness—the very courses of progress—may be less than benign. Scientific progress may pose the problem of immoral means used for moral ends. But the even more important question concerns the ends of scientific progress themselves. Are there secrets about Bensalem that remain after the sailors' "discovery" of Bensalem to the rest of the world? If so, then *New Atlantis* may be incomplete after all, and this because of problems inherent in the world's scientific destiny.

The history of Bensalem, related by the priest, answers the sailors' questions about the island's conversion to Christianity and about why the island, at least some of whose inhabitants know all about the outside world, is unknown to that world. This latter odd condition is caused by the Bensalemite policy concerning the admission of strangers and by the laws of secrecy governing the few Bensalemites who travel abroad.

Bensalem's conversion to Christianity appears to be the result of a miracle of original revelation. But again the appearance fades with closer inspection. For according to the priest's report, the miracle was by chance confirmed to be genuine by

[7]*Above*, pp. 72, 76–78, 82–86.

one of the "wise men" of Salomon's House—a natural scientist—whose institution has been "vouchsafed" by God to "to know thy works of creation, and the secrets of them; and to discern (as far as appertaineth to the generations of men) between divine miracles, works of nature, works of art, and impostures and illusions of all sorts."[8] The whole story thus introduces the issue of miracles and frauds, but it really leaves the matter unsettled because the scientists know the secrets of nature only as far as appertains to the "generations of men." This event occurred "about twenty years after the ascension of our Savior." As we'll see below, that means roughly *sixteen hundred* years before the report by the priest to the sailors. So there is no way the scientist could have at the time of the revelation been sure that there was no natural or artificial cause of the divine occurrence.

It is useful in this context to refer to the treatment of miracles in the *Advancement of Learning*, where Bacon complains that ecclesiastical history has too easily accepted reports of miracles by holy men, which were believed for a time because of ignorance and superstitious simplicity but then with the passage of time as "the mist began to clear" became "esteemed but as old wives' fables and impostures of the clergy." He then praises Aristotle for having, in his natural history, distinguished between matters of manifest truth and seemingly incredible matters of prodigious narrations and rarities. But he also praises Aristotle for not suppressing the incredible matters and for not denying them to the memory of men.[9] That's of course because some of these incredible and rare natural things might prove to be true. But why not also for the wives' fables or impostures of the clergy, even though the mist has arisen and they are "esteemed" (not "known") to be false? Bacon doesn't tell us what it means for the mist to arise. And why could it not arise as to the miracle supposedly verified by the wise man of Salomon's House? It might be true; but then

[8]*Above*, pp. 73–76.
[9]*Advancement of Learning*, hereafter *Advancement*, BW, III, 287–89.

again it might not. But from what Bacon tells us from the priest's speech, the odds seem to tilt toward the whole thing being, if not a wives' tale, an imposture of the clergy, or rather an imposture of the House of Salomon. Even if the priest believes all this, whoever long ago first told of the miracle could have made it all up. It doesn't matter if the miracle was reported as seen by a crowd rather than an isolated individual, like Moses. A crowd was said to have seen the miracle of the loaves and fishes. In the matter of miracles there is always room for doubt or fraud.

This doubt is magnified by the striking fact that the central event in Bensalem's religious history is the miracle of revelation and that a scientist verifies it, on the one hand; and that the scientists have mastered all the techniques of art and illusion and have "so many things truly natural which induce admiration, could in a world of particulars deceive the senses, if we would disguise those things and labour to make them seem more miraculous."[10] The Father of Salomon's House says the fellows are forbidden to engage in such fakery; but then again, the fellows "take an oath of secrecy, for the concealing" of things they do not want to reveal. It is hard not to suspect that Bensalem's Christianity is the product of *scientific* art and illusion.

The priest's speeches in the sequel present the civil history of Bensalem, in order to explain to the sailors something of why the extraordinary island is unknown to the rest of the world. In fact, the history recounts the coming of science and scientific rule to Bensalem. And the most impressive aspect of this history, not even discussed in the story, is that for Bensalem, unlike the rest of the world, the advent of science preceded the coming of Christianity. We know this from the obvious fact that a wise man of Salomon's House, the scientific establishment founded by King Solamona, verified the miracle revealing Christianity to the island, and from the less obvious but still clear chronological evidence from the story: If the "six score years" since navigation had begun to increase, mentioned by

[10]*Above*, p. 108.

the priest, be reckoned from the obvious beginning of European navigation, 1492, the sailors' voyage took place in 1612.[11] Therefore, the time referred to by the priest as "three thousand years ago," when the great Atlantis "did flourish," would be 1388 B.C.E., 458 years before the biblical Solomon completed the temple in 930 B.C.E., and the time of King Solamona's rule in Bensalem, "about nineteen hundred years ago," would be 288 B.C.E.[12] In other words, the coming of science to Bensalem preceded the coming of Jesus to the world by almost three hundred years. Given this priority, we might wonder if somehow the advent of the Christian world needed Bensalemite science. But at the same time we're induced to wonder why science, as it appeared in Bensalem, needed Christianity.

We can follow up these questions by considering the *Advancement of Learning*, where Bacon has some interesting things to say about the order of historical destiny. There he says that in the modern age men will imitate the heavens by circumnavigating the globe "as the heavenly bodies do" and that God has ordained that such circumnavigation will be coeval with "the further proficience and augmentation of all sciences."[13] The Bensalemites engage in such circumnavigation, and the narrator, to whom the wonders of Bensalemite science have been revealed, writes in English to an English audience, as if Great Britain were to lead the world to its Bensalemite destiny. The British—*the modern seafaring people*—are in fact the people who will

---

[11]*See* Howard B. White, *Peace Among the Willows: The Political Philosophy of Francis Bacon* (The Hague: Martinus Nijhoff, 1968), pp. 104 n. 30, 121–22. The 1612 date puts the voyage after the publication of the *Advancement of Learning* and before the *Novum organum* and *De augmentis*, and it provides a clue to the importance of the *Advancement of Learning* in the whole of Bacon's writings. The division of the sciences contained in the *Advancement of Learning* provides the description of the intellectual globe that will tame and conquer the terrestrial globe. Cf. White, pp. 104 n. 30, 121–22, 197–98.

[12]*Above*, pp. 78, 82.

[13]*Advancement*, BW, III, 340.

become Bensalem when guided by Bacon's instauration. No wonder, then, that the crucial period of British history, the period from "uniting of the Roses to the uniting of the Kingdoms" (from Henry VII's founding of the modern British state to the combining of the English and Scottish monarchies— one hundred years before actual political union—with the accession of James I), is "after the manner of the sacred history, which draweth down the story of the Ten Tribes and of the Two Tribes as twins together."[14] In this age the modern state was built and Bensalem—modern science—was revealed to it by Bacon's great instauration. But if Britain is now joined in monarchy "for ages to come," and if British and Bensalemite destiny are the same, ushering in a wholly new age for mankind, it is surprising that in mentioning the likeness of British history to the sacred history Bacon is silent about the coming of Jesus: he speaks only of the sacred history before the coming of Jesus, in particular the sacred history from the Exodus to the division of David's empire.[15] Not only in Bensalem, then, but in the world that would receive Bensalemite science, the true order of things is for Christianity to appear after modern science. Surely this must perplex us, unless we assume that modern science is not really different from Christianity—that science is but a mode of Christian faith. Only with such an assumption does Bacon's story not reverse the actual historical order. And with this assumption the Christianity introduced to the Bensalemites would have been but a new version of their old and established science, which had been all along a form of charitable belief.

But then what are we to make of the divine ordination, after the circumnavigation of the globe, of further advancement of

---

[14]*Ibid.*, p. 336.
[15]*Bacon* wrote a history of the reign of Henry VII, BW, VI, 23–245. This work is important for understanding Bacon's understanding of the relation between science and the modern state. It will be interesting for the reader to compare the account of Henry's statesmanship with that of King David, whose kingdom was doomed by the prophecy of Nathan. See above, p. 124 note 11.

all sciences in an England whose history is like the story of the ten tribes and the two tribes as twins together? It seems to have no place for Jesus. According to the clues about historical destiny provided by the *Advancement of Learning*, the true beginning of the new age—the concatenation of modern Britain and Bensalemite (Bacon's) new science—is better likened to David's empire than to the kingdom of Jesus. It seems, then, that however Christian it may be, Bensalem's destiny is unaffected by the difference between David and Jesus. From the standpoint of pious Christianity, which certainly characterizes the outward appearance of Bensalem, this difference is of paramount importance. Thomas Jefferson called Bacon "one of the three greatest men to have ever lived, without exception" (along with Isaac Newton and John Locke) "and as having laid the foundations of those superstructures which have been raised in the physical and moral sciences." We should not be surprised that the Baconian Jefferson assembled a version of the Gospels that included no miracles, including the resurrection of its central figure.[16]

We have already come to suspect that Bensalemite hospitality and humanity, presented in the guise of Christian charity, are something more than first meets the eye. And one cannot but notice that the character who tells the narrator about Bensalem's Christian sexual morality is Joabin, namesake of the biblical Joab, who we know, of course, assisted David's sin against Uriah the Hittite, for which the rule of David's successors was cursed by Nathan's prophecy. It is also strange that the unit of linear measure in Bensalem is the "karan," a word whose Hebrew origin means "horn" and "ruler of the House of David."[17] All of this suggests that Bensalem, the true destiny of Christianity, is indifferent to the distinction between the old and the new covenants, which according to Christian doctrine

---

[16]*Letter* to John Trumbull (February 15, 1789). Thomas Jefferson, *Writings* (The Library of America, 1984), pp. 939–40. See *The Jefferson Bible* (Washington, D.C.: Smithsonian Books, 2011).

[17]*Ezekiel* 29: 21; Psalms 132: 17. The verb form means to send out rays or display horns.

consists (among other things) in the saving grace of Jesus, whose divine perfection contrasts with the sinful weakness of King David, the very first of the House of David. Before we can follow up these clues, which are developed only in the account of Bensalem's present, we must return to the account of the island's history.

To explain why Bensalem is unknown to the rest of the world, the priest relates first the story of the ancient Bensalemites' encounter with the old Atlantis, the one the priest calls "the great Atlantis" and the sailors "call America," and then tells the story of the lawgiver Solamona, who brought science to what has become the New Atlantis. The first story explains why the world no longer comes to Bensalem, as it did "about three thousand years ago," and the second explains why the Bensalemites do not travel freely to the rest of the world, as they did before the reign of Solamona "about nineteen hundred years ago."[18]

Now according to the first story, the rest of the world once travelled often to Bensalem and navigation was then "greater than at this day." But the navigation to Bensalem from America ultimately disappeared, because of the divine revenge—in the form of a flood—that swept away the "proud enterprises" of the old Atlantians. And the navigation to Bensalem from other parts of the world disappeared because of wars and the "natural revolution of time."[19] The proud enterprises of which the priest speaks refer to the conflict between the ancient Bensalemites under King Altabin, and Peru, and Mexico, which along with the great Atlantis were "mighty and proud kingdoms in arms, shipping, and riches." In the space of ten years Mexico attacked Europe and Peru attacked Bensalem. Plato, the "great man with you" mentioned by the priest, must therefore have been wrong, in his account of Atlantis, that the attackers of Europe were the ancient or great Atlantians, who must rather have minded their own business. And this is not all

[18]*Above*, pp. 78, 82.
[19]*Above*, pp. 79–82.

about which Plato was wrong, as closer inspection of the story reveals.

Contrary to Plato's account the priest says that the Athenians—if indeed they were even the ones who saved Europe—were no less ruthless than their Mexican attackers, as appears from the accounts of the fate of the invaders of Europe at the hands of the Athenians: no man or ship returned to Mexico. But the Peruvians would have suffered the same fate if it had not been for the bloodless victory and extraordinary clemency of Bensalem's King Altabin. And while in Plato's account it is unclear whether, after the great war, Atlantis was destroyed by natural disaster or divine revenge (Zeus' righteous chastisement), the priest leaves no doubt at all. What he first calls "divine revenge" as the reason for the great Atlantis's destruction, he later calls quite simply a "main accident of time."[20] At any rate, in Plato's account earthquakes and floods soon after the Athenian victory destroyed the entire body of Athenian warriors and caused Atlantis to be swallowed by the sea. And the deeds did not live even in Athenian memory because Athens, along with the rest of the world except for Egypt, was subject to "many and diverse destructions of mankind, of which the greatest are by fire and water, and lesser by countless other means."[21] In Plato's account, Egyptian antiquity—and so the surviving memory of the Athenian deeds—depends on the Egyptians' piety, for although the Egyptians are immune from destructions caused by rain and fire, because of the fortuitous presence of the Nile, they are not immune from earthquakes.[22] According to the priest the Egyptians' antiquity was more likely due to natural chance: he is careful to note that the "whole tract" referred to by Plato "is little subject to earthquakes."[23]

---

[20] *Ibid.*

[21] *Timaeus* 21c1–23d1.

[22] *Seth* Benardete, "On Plato's *Timaeus* and Timaeus' Science Fiction," *Interpretation*, 2 (Summer, 1971), p. 30; *Timaeus* 22c1–23a1, 25e8–d7.

[23] *Above*, p. 80.

In other words, Plato was wrong to suggest the influence of divine revenge and Egyptian piety. Rather, between the time of Altabin and the time of King Solamona both Egypt and Bensalem survived by sheer good fortune the world's accidental destruction. After Solamona's reign, however, the Egyptians' luck ran out and they were among the peoples who once sailed to Bensalem but who have not done so since the ages after Altabin's time.[24] And after Solamona's reign, the Bensalemites used the powers of their natural science to withstand the natural revolutions of time. The Bensalemites did not at first need piety, because their king Altabin was too humane to have called forth divine revenge. So why did the island need Christianity? Perhaps to absolve the Bensalemites of guilt for (unmentioned) crimes committed before the reign of Altabin, or perhaps to absolve them of guilt for crimes committed in the pursuit of science: we know that Solamona's laws of secrecy may be far less humane than Altabin's deeds. But however deserved some divine revenge may have been, after the reign of Solamona, whose intention was "to give perpetuity to that which was in his time so happily established," the Bensalemites need not have feared any divine punishments of the kinds suffered by the rest of the world.[25] We must not forget that the new science heralded by the Bensalemites does not blush at aiming to overcome mortality, the very wages of sin. Even more we are forced to wonder why the island needed Christianity. Is Bensalem's natural science somehow unable to provide its own moral justification? Or perhaps it never really mattered.

After hearing the stories of Altabin and Solamona the sailors declare themselves to be free men. No longer fearing the danger of their "utter perdition," they find the Bensalemites' hospitality toward strangers to be so free as to incline them to forget

[24]*Above*, p. 78.
[25]*Above*, pp. 82–83.

what was dear at home.[26] The nature of this freedom—both the strangers' and the Bensalemites'—is characterized in the speeches about Bensalem's present, of which the first describes Bensalem's ultimate promise and the second describes the regulation of that promise. The first speech is made by the narrator and describes the "Feast of the Family." This feast honors any Bensalemite man "that shall live to see thirty persons descended of his body alive together, and all above three years old."[27] The honor recognizes mere longevity and fecundity, not any excellence of soul. For while the state executes the orders of the father, who is called the "Tirsan," to mend the vicious ways of any members of his family, there is no mention of any moral requirements to be met by the father himself. And yet according to the narrator the feast shows Bensalem to be "compounded of all goodness."[28] Bensalem, described later as "the virgin of the world,"[29] is in fact dedicated to the preservation and generation of human bodies, and thus also to the most intense pleasure that accompanies the generation. The full meaning of this fact is disclosed by the second speech, made by the character named Joabin.

Prompted by having heard of the ceremony of the Tirsan, in which "nature did so much preside" and which concerns the

[26]*Above*, p. 87. The sailors declare their liberation in the sixteenth paragraph of the story. The sixteenth essay in the *Essays* treats atheism and is followed by the seventeenth essay on superstition. The principle of the sixteenth essay is that any belief about God is better than unbelief. The principle of the seventeenth essay is that unbelief is better than wrong or heretical belief. Given that such a contradiction must always redound to the benefit of unbelief or atheism—no believer will *ever* pronounce the truth of atheism—Bacon in the *Essays* makes an argument for atheism. In the seventeenth paragraph the narrator forgets a difference between six and seven days. If essays sixteen and seventeen come together as a true essay on atheism, the number of essays is equal to the number of paragraphs in *New Atlantis*.

[27]*Above*, pp. 87–91.

[28]*Above*, p. 87.

[29]*Above*, p. 93.

propagation of families, the narrator asks Joabin about "nuptial copulation" in Bensalem: about the laws and customs concerning marriage and especially about whether the Bensalemites practice polygamy. Joabin's answer consists of a long speech about European and Bensalemite sexual morality and a brief mention of the "Adam and Eve's pools."[30]

According to Joabin, there could be no greater contrast between the chaste Bensalemites and the corrupt Europeans. Whereas Bensalem is the most chaste of nations, "free from all pollution and foulness" and "the virgin of the world," European sexual practices are debauched. The Bensalemites view marriage as the cure for "unlawful concupiscence" and think of natural concupiscence as "a spur to marriage." The Europeans—with their brothels, baths, infidelities, and "masculine love"—have put marriage "out of office." Even those Europeans who do marry continue in their debauched ways and justify them as done to prevent greater evils, such as adultery, the deflowering of virgins, and "unnatural lust." This, says Joabin, the Bensalemites consider a preposterous wisdom which they call "Lot's Offer." At this point the narrator interrupts to comment, as the widow of Sarepta did to Elias, that Joabin had come to remind the narrator of his sins. It is true, admits the narrator, that the Bensalemites are indeed more righteous than the Europeans. Joabin then continues by outlining the laws that govern marriage in Bensalem: there is no polygamy; there is a month wait between first interview and marriage or contract for marriage; marriage without parental consent is mulcted "in the inheritors" (i.e., the offspring of the marriage are penalized); and finally, "because of many hidden defects in men and women's bodies," before a contract is made one friend of the potential bride and one friend of the potential groom are permitted to see the bride and groom "severally bathe naked" in the "Adam and Eve's pools" that are situated near every town.

Joabin's speech is the most important in *New Atlantis*. Except for the scientist who judged the miracle, Joabin is

[30]*Above*, pp. 91–95.

the only post-Solamonic character described as wise. Not even the Father of Salomon's House who makes the story's final speech is described as wise. Moreover, not only is Joabin wise, but he is said to be "learned . . . of great policy, and excellently seen in the laws and customs" of Bensalem. Joabin is wise in matters of government and rule, and as such he espouses his Jewish doctrines, and their compatibility with Bensalemite Christianity, from considerations of policy rather from any conviction of truth.[31] As a man wise in general and wise in matters of policy, Joabin transcends the myths of Bensalem. Of all the characters in the story, only Joabin's wisdom, though apparently not scientific, is otherwise comprehensive, and in this respect it certainly matches, if not surpasses, the wisdom of the large-hearted founder and lawgiver Solamona. It is all the more important, then, that it is Joabin whose speeches disclose the full nature of Bensalem's corporeal promise to humanity.

As Bacon has Joabin report it, the Bensalemites' likening of European opinions about sex and marriage to the preposterous wisdom of "Lot's Offer" is subtly inappropriate. Joabin argues that the Europeans claim to offer daughters (as prostitute-wives) to save other daughters, the potential victims of unnatural lust and rape. But Lot offered his virgin daughters to save male strangers, an act of extraordinary hospitality.[32] Bacon has Joabin express Bensalemite abhorrence of the offering of daughters to save daughters, while pointing to and remaining silent about the practice of offering daughters for the sake of visiting male strangers. This fact should certainly cause us to wonder about the priest's earlier remark that the sailors should feel free to make "any other request" of the Bensalemites, which would not be refused, and to note Bacon's

---

[31]*The* narrator says of Joabin only that "surely this man . . . would ever acknowledge" the basic reality of Jesus, not that he believed in it, and of the "Jewish dreams" that must be set aside in order to count Joabin wise the narrator says only that Joabin was desirous "by tradition among the Jews there" to have them believed, not that Joabin believed them himself. Above, pp. 91–92.

[32]*Genesis* 19: 1–11.

silence about what is commonly thought to be the most pressing need felt by sailors who are newly in port.[33] And when the narrator says that Joabin's speech made him feel like the widow of Sarepta we are reminded again of the harsh possibilities lurking behind Bensalem's outward hospitality. In the Bible the widow commented not just that Elias had come to remind her of her sin, but also that she feared him to have come to kill her son.[34]

These facts lead us to wonder about the extraordinary institution of the Adam and Eve's pools. This institution is intended to regulate nuptial copulation by preventing the disappointments—and by implication the infidelities—that might spring from the discovery after marriage of the "many hidden defects in men and women's bodies."[35] Joabin says that because it involves only indirect familiarity by way of friends, the Bensalemites' version of naked prenuptial interview is superior to the one devised by "one of your men," who included it in an account of a feigned commonwealth. It is not clear whether by his remark Joabin refers to Plato's *Laws* or More's *Utopia*. But in either case it would seem that the Bensalemites' practice is in fact the least workable. In Plato's Magnesia boys and girls see each other frequently and in common in naked play, and naked only within the limits set by a moderate sense of shame, and in Utopia the spouse is seen by the prospective spouse. In Bensalem, however, the prospective spouse is viewed by a friend of the other partner.[36] What, we are tempted to ask, can prevent a friend from making a false report of bodily defect in order to secure the intended partner for himself or herself, a result entirely possible since Joabin does not specify the sex of the reporting partner? And thus what is to prevent adulteries inflamed by the desire of friends for the husbands and wives of friends, desire grounded in familiar knowledge of bodies?

[33]*Above*, pp. 70–71.
[34]*I Kings* 17: 18.
[35]*Above*, p. 95.
[36]*Plato, Laws* 771e; More, *Utopia*, "Their Marriage Customs."

According to Joabin the Bensalemites are restrained from vice by religion and by "reverence of a man's self"; that is, by conscience and self-love.[37] But Bensalemite conscience is in fact not entirely scrupulous, as we know from having examined their apparent charity and hospitality. We cannot be confident that the Bensalemites' self-love is fully armored against mere selfishness. And just as these questions and doubts arise, Joabin's remarks are interrupted so that we hear no more about the matter, our attention being turned instead to the disclosure of Salomon's House. The institution of the pools presumes that the Bensalemites can be moved by the desire for other, more beautiful bodies—why else would they need to prevent fraud and disappointment? But as presented the institution points more to problems than to sweet marital bliss. It points to the licentious powers of choosiness, the love of one's own, and the desire for more, which could reside at the very heart of Bensalemite "goodness."

The problems of Bensalem's erotic practice are highlighted by what is by now surely obvious: The only Bensalemite said to be wise in general and wise in policy, and who reveals the astonishing practice of the pools, is the namesake of the biblical Joab, David's strong lieutenant in his rise to power and one of the most ruthless men in the Old Testament. Joab treacherously assassinated Abner and Amassa, murdered David's son Absalom, and carried out David's treachery against Uriah the Hittite. Joab assisted David in the theft of Uriah's wife, and her adultery, that resulted from David's having glimpsed Bathsheeba bathing naked. Then at David's orders, he arranged for the virtual murder of the loyal Uriah before the gates of Thebez. It was for this injustice that the Prophet Nathan cursed David and his house.[38] It is almost a joke for Bacon to have Joabin be the one to tell the narrator, and us, about the Bensalemite practice of having individuals see their friends' prospective spouses "bathe naked."

[37] *Above*, pp. 94–95.
[38] *II* Samuel 2: 8–II Samuel 20: 10; I Kings 2: 5–46.

But why would Bacon tell such a joke? Perhaps to spur us to think about why the Bensalemites seem so creepy. When the Bensalemites with truncheons first size up the sailors and forbid them to land (for the sake of public health, it turns out), they display no fierceness. And the Bensalemite who boards the ship does so with no distrust.[39] When the Bensalemite officer leads a small party of the sailors to see the Strangers' House, there were Bensalemites lining both sides of the fair streets all "standing in a row," in "so civil a fashion, as if it had been not to wonder at us but to welcome us," and they made welcoming gestures.[40] The Governor-Priest of the House of Strangers weeps tears of tenderness that cause the sailors to think they had come to a land of angels.[41] The Bensalemites are so humane that the sailors forget about what they held dear in their own countries.[42] The eastern city where the miracle occurred is called Renfusa, a combination of Greek terms that mean "sheep-natured."[43] That term could well be used to describe the Bensalemite's demeanor: in their outward behavior they seem completely composed and orderly. All present at the Feast of the Tirsan proclaim in unison: "Happy are the people of Bensalem."[44] The offspring of the Tirsan almost never disobey.[45] And the name Tirsan comes from a Persian word for "timid" or "fearful." When the Father of Salomon's House enters the city the people line up like soldiers in battle array and they stand in windows "as if they had been placed." And the Father has an aspect "as if he pitied men."[46] The Bensalemites appear to be a gentle flock attended at once by Joabin and by the scientific establishment, which possesses powerful canons and weapons and to which even the state is subservient since

[39]*Above*, p. 64.
[40]*Above*, p. 67.
[41]*Above*, p. 72.
[42]*Above*, p. 87.
[43]*Above*, p. 73.
[44]*Above*, p. 90.
[45]*Above*, p. 88.
[46]*Above*, pp. 87, 97, 96.

the scientists decide what to reveal to the state and what not to reveal. But if in fact the Bensalemites are really no different from the likes of David and Joab, then we have to wonder how they remain so tame and calm. One thought immediately comes to mind: the House of Salomon produces all kinds of unheard of medicines and it is not impossible that one of them is a narcotic the Bensalemites take, or are fed, every day. That possibility would be compatible with the ominous currents we have seen lurking in Basalem's smiling hospitality.

Rawley's claim that Bacon failed to describe the best form of commonwealth is misleading. In his essay "Of the True Greatness of Kingdoms and Estates" Bacon addressed exemplary empires and in his 1622 *The History of the Reign of King Henry the Seventh* he described at length an exemplary prince. The lesson of the essay is simple: Statesmen who want to make a state great must focus on one thing: war. The basic task of the statesman is thus to ensure the warlike character of the citizenry. Money is not, says Bacon, the sinew of war "where the sinews of men's arms, in base and effeminate people, are failing."[47] The primary means to this end are to honor war, to allow for naturalization of foreigners, to avoid burdensome taxation, to avoid dainty indoor arts, to have policies and laws that provide moral pretext (however specious they may be) for foreign war, and especially to prevent the predominance of a nobility and gentlemen and at the same time develop a large class of "middle people" with enough land to be independent and thus "fit for an helmet." Bacon's example of the latter is none other than "king Henry the Seventh," whose policy in this regard was "profound and admirable." Henry made "farms and houses of husbandry of a standard; that is, maintained with such a proportion of land unto them, as may breed a subject to live in convenient plenty and no servile condition; and to keep the plough in the hands of the owners and not mere hirelings."[48]

[47]*Above*, p. 123.
[48]*Above*, p. 124.

According to Bacon's history, Henry VII at the end of his reign was as at "the top of all worldly bliss, in regard of the high marriages of his children, his great renown throughout Europe, and his scarce credible riches, and the perpetual constancy of his prosperous success."[49] He was, then, indeed an exemplary prince and human being as well. But according to the history, Henry had no stomach for war at all, despite his policies of reducing the nobility and empowering a free and yeoman middle class. Moreover, his reign was marked by his rapacious greed and duplicitousness: rather than make foreign war he used the threat of it, and then made dishonorable peace, to bilk parliament and the nobility of huge sums. He used illegitimate "legal" prosecutions as well to fleece the nobility. And even though his policies enhanced the political power of the people, he taxed them so much that they came to hate him. He was, however, concerned with matters of civil war, which, in "Of the True Greatness of Kingdoms and Estates," Bacon called "the heat of a fever" and unlike foreign war that is "like the heat of exercise, and serveth to keep the body in health."[50]

Henry came to power to end the three-decade civil wars known as the Wars of the Roses, fueled by rival claimants (the noble Houses of York and Lancaster) to the English throne. As always, these partisan quarrels among the great, aristocratic houses meant murder and rapine for the people. Henry did not blush from letting it be thought by the people that he had been ordained by heaven to put an end to their misery. But at the same time, throughout most of his reign he not only kept track of, but even encouraged before crushing, partisan conspiracies against him. He was never in real danger; but his policy was devilishly clever: he used spies and planted agents in the conspiracies, and profaned religious oaths and violated the sanctity of the confessional to gain information, and in crushing the conspiracies used great executions and great pardons, all as means to his victorious ends. The result was to remind the

---

[49] See *The History of the Reign of King Henry the Seventh*, ed. Jerry Weinberger (Ithaca: Cornell University Press, 1996), p. 202.
[50] *Above*, p. 128.

people of how much they needed him, despite his loathsome taxes, to keep what money they had left and not be tormented by ambitious partisans. All of the conspiracies in the end failed because they lacked sufficient popular support. Henry built a great monarchy, but his popular policy and contempt for nobility and honor paved the way for a polity in which kings are gone or merely ornamental.

In *The History of the Reign of King Henry the Seventh*, Bacon presents his understanding of the modern state: Its end should be money for the state and for the people as far as possible, not war and empire. Political partisanship is always delusional and the end of political progress should be to make it disappear. (This is why in "Of the True Greatness of Kingdoms and Estates" Bacon condemns the "wars which were anciently made on behalf of a kind of party . . . or to set up or pull down democracies or oligarchies."[51]) The state's concern is the human body, not the soul. Its principle should be that all people want the same thing, what Bacon's student Thomas Hobbes took to be long life and "commodious living," and that means money and good health.[52] These were ultimately Henry's concerns, not justice, party, or honor. And these are what the scientific conquest of nature can provide to a degree that even Henry could never have imagined.

New Atlantis ends with the magnificent return of a Father of Salomon's House, his revelation to the narrator of the "true state" of the scientific establishment, and the Father's gift of a "great largess" or tip to the narrator and his fellows.[53] The Father's description of the scientific establishment reveals the institutional means for Bacon's Great Instauration and the conquest of nature. One field of study, however, is conspicuously absent: political science. It seems reasonable to think that, for Bacon, in the new scientific world to come there simply isn't anything we need to know about politics. The Bensalemites seem

[51]*Ibid.*
[52]*Hobbes*, *Leviathan*, ed. Richard Tuck (Cambridge: Cambridge University Press, 1996), p. 90.
[53]*Above*, pp. 95–111.

perfectly contented and there is no mention of justice or partisanship as being among their concerns. Indeed the words "justice" and "party" *never* occur in *New Atlantis*.[54] That said, Bensalem still has a king and Joabin is wise about its ancient laws, which his fellow Jews believe came from Moses. Whatever those laws are, however, it seems that the scientists call the ultimate shots (what to reveal and what to hide from the state) from behind a veil of secrecy: they must think that some things about their government are not fit to be uttered. One candidate could well be that the miracle of Christian revelation was a scientific fraud. If that were hidden in the annals of Salomon's house the scientists and only they would know for sure that indeed it was.

But what if the Jews are right and Moses really was the source of Benslemite law? Bacon has his narrator glibly dismiss this as a "dream" and thus not a miracle from God.[55] But how would he or any of the scientists know this? In *New Atlantis*, Bacon takes us in two theological directions. On the one hand, he pushes us hard toward thinking that the miracle of revelation in Bensalem was a fake, just as all such things are, only in Bensalem's case it really looked like a miracle because the scientists, unlike monks and the clergy, really can manufacture something that looks just like one. Science is science and everything else is superstition. But on the other hand, our skepticism is fueled by the fact that the scientist who verified the miracle could not have really done so demonstrably because his science was almost sixteen centuries out of date. But that fact then leads us to wonder: what if science can be fully completed and there is an end to the generations of men who make further progress? Bacon seems to suggest that *all* the latent structures and processes of nature can in principle be known. In that case we really would have to admit that an

---

[54]*The* word "honor" is used seven times, but in every case it indicates a sign of esteem or reward for some service or accomplishment. It never refers to a moral obligation performed without regard for consequences. See above, pp. 61, 67, 85, 89, 90, 98, 110; cf. p. 72.

[55]*Above*, p. 92.

occurrence outside of those structures would have to be miraculous. And such an occurrence could well be a scientifically inexplicable voice of conscience within.[56]

By the end of *New Atlantis* we're still left with a puzzle: can human beings really live by reason and natural science alone, or do they inevitably need, or at least yearn for, the rule of God? If the answer is the latter, as Bacon seems to concede in *New Atlantis*, then a significant problem could exist in Bensalem. In his essay "Of Unity in Religion," Bacon concedes that the ancients got one thing right: "The quarrels and divisions about religion were evils unknown to the heathen." This is because their religion "consisted rather in rites and ceremonies, than in any constant belief" and was concocted by their poets. "But the true God has this attribute, that he is a *jealous God*; and therefore his worship and religion will endure no mixture nor partner."[57] As a consequence of this rigidity, doctrinal innovation becomes especially dangerous for the religion of the Book. With every doctrinal split comes two opposed but absolutist and fanatical religious parties and consequent "cruel and execrable actions of muthering princes, butchery of people, and subversion of states and governments."[58] And one schism always leads to others. Bacon refers to Lucretius' comment about Agamemnon's sacrifice of his own daughter: *Tantum Relligio potuit suadere malorum* [to such ill actions Religion could persuade a man].[59] But this, says Bacon, is nothing by comparison to what goes on in Christendom. Had Lucretius known of such things as the St. Bartholomew's Day massacre or the gunpowder plot in England he would have

---

[56]**For** further discussion of this question see my "On the Miracles in Bacon's New Atlantis" in *Francis Bacon's* New Atlantis: *New Interdisciplinary Essays*, ed. Bronwen Price (New York: Manchester University Press, 2002), pp. 106–28; and "Francis Bacon and the Unity of Knowledge: Reason and Revelation" in *Francis Bacon and the Refiguring of Early Modern Thought* (Burlington: Ashgate, 2005), pp. 109–27.

[57]*Above*, p. 113.

[58]*Above*, p. 117.

[59]*Above*, pp. 116–17.

been "seven times more Epicure and atheist than he was." The temporal sword is to be used with great care in matters of religion and it is "monstrous," he continues, to put it into the hands of the common people as do the Anabaptists and "other furies."[60]

Theological innovation is the poison peculiar to Christianity, which makes doctrinal unity essential. The Bensalemites are Christian and at the time of the miracle there were Hebrews, Persians, and Indians and all miraculously were able to read the Bible in their own languages and were thus saved from infidelity. We learn that there are still in Bensalem lineages and tribes of Chaldeans and Arabs and others who came when the rest of the world travelled to its shores. We don't know if they are Christians, although we know there are Jews there who acknowledge the divinity of Jesus. Bensalem appears to be a tolerant place with simple and easy doctrines. But how can the dangers of religious bloodshed be avoided among a people so used to the one thing Bacon tells us makes the religion of the Book so prone to them: innovation? Might not a people used to ever new wonders of science also tend to want new wonders in spiritual matters as well? Bacon doesn't say, and he seems to think that the Bensalemites' material plenty just makes them spiritually tame. But if he's wrong, then for sure Bensalem would need unity of religion and Bacon does not tell us who would preserve it or how it might be done.[61]

Thomas Jefferson was right to include Francis Bacon in his triumvirate of founders of the modern physical and moral sciences. Bacon was far-sighted in his understanding of modernity. First and foremost, he understood that the ultimate

---

[60] *Ibid.*

[61] *See* Travis DeCook, "Francis Bacon's 'Jewish Dreams': The Specter of the Millennium in *New Atlantis*," *Studies in Philology* 110.1 (2013). DeCook argues persuasively that Bacon had matters of religious unity in mind, especially since he had participated in the 1618 trial of John Traske for religious heterodoxy in claiming that Christians should celebrate the Sabbath on Saturday and obey Jewish dietary laws.

ground of modernity would be the scientific and technological conquest of nature. He likewise understood that a certain kind of politics would be needed to bring that scientific and technological project into being: the politics he outlined in his *History of the Reign of King Henry the Seventh*. But in addition to these powerful insights, *New Atlantis* discloses that Bacon seems also to have envisaged the discontents and problems of the modern technological age. Can material plenty do away with pride and the desire for distinction and conflict over rare objects of desire? Does technological modernity somehow diminish the human soul (if there is one)? And most important, can technological modernity eradicate the human longing for God and divine law. And finally, can technological modernity eradicate the voices of conscience that, if Bacon is right, inevitably come into conflict? Francis Bacon was not only one of the founders of the modern age. He was also its first and perhaps its most thoughtful diagnostician.

Bibliography

Biographies

Aubrey, John. *Brief Lives*. Edited from the original mss. and with a life of John Aubrey by Oliver Lawson Dick; foreword by Edmund Wilson. Ann Arbor: University of Michigan Press, 1962.

Bowen, Catherine Drinker. *Francis Bacon: The Temper of a Man*. Boston: Little, Brown and Co., 1963.

Church, R. W. *Francis Bacon*. Published in 1884. Nisyros Publishers, 2015.

Jardine, Lisa and Alan Stewart. *Hostage to Fortune: The Troubled Life of Francis Bacon*. New York: Hill and Wang, 1998.

Macaulay, T. B. *Critical and Historical Essays* Vol. 2: "Lord Bacon" (July 1837). Online Library of Liberty.

Sturt, Mary. *Francis Bacon*. New York: William Morrow and Co., 1932.

Bacon's Works

The standard, long-definitive edition of Bacon's works is by Spedding, Ellis, and Heath, 14 Vols. (London: 1857–1874). This includes Rawley's *The Life of the Right Honorable Francis Bacon*. This edition is reprinted in The Cambridge Library Collection, Cambridge University Press.

Translations of *Temporis partus masculus* (*The Masculine Birth of Time*), *Cogitata et visa* (*Thoughts and Conclusions*), and

New Atlantis and The Great Instauration, Second Edition.
Edited by Jerry Weinberger.
© 2017 John Wiley & Sons, Inc. Published 2017 by John Wiley & Sons, Inc.

Redargutio philosophiarum (*The Refutation of Philosophies*) can be found in Benjamin Farrington, *The Philosophy of Francis Bacon*. Liverpool: Liverpool University Press, 1964.
A new and certain-to-be definitive edition of Bacon's works is underway in the Oxford Francis Bacon Project, published by Oxford University Press. To date seven of the planned 15 volumes have been published, including the wonderful edition of the 1620 *Novum organum* by the late Graham Rees with Maria Wakely: *The Oxford Francis Bacon*, Vol. XI. Oxford: Clarendon Press, 2004. Volume VI contains translations of *Phænomena universi* (*Phenomena of the Universe*), *De fluxu et refluxu maris* (*On the Ebb and Flow of the Sea*), *Descriptio globi intellectualis* (*A Description of the Intellectual Globe*), *Thema cœli* (*Theory of the Heaven*), *De principiis atque originibus* (*On Principles and Origins*), and *De vijs mortis* (*An Inquiry Concerning the Ways of Death*). Volume XII contains translations of Bacon's *Historia naturalis et experimentalis* (*Natural and Experimental History*), *Historia ventorum* (*History of the Winds*), and *Historia vitæ & mortis* (*The History of Life and Death*). Volume XIII contains translations of *Historia densi & rari* (*History of Dense and Rare*), *Abecedarium nouum naturæ* (*A New Abecedarium of Nature*), *Historia & inquisitio de animato & inanimato* (*History and Inquiry Concerning Animate and Inanimate*), *Inquisitio de magnete* (*An Inquiry Concerning the Loadstone*), *Topica inquisitionis de luce et lumine* (*Topics of Inquiry Concerning Light and Lumen*), and *Prodromi sive anticipationis philosophiæ secundæ* (*Precursors Or Anticipations of the Philosophy to Come*).
Francis Bacon. *Novum Organum: With Other Parts of The Great Instauration*. Translated and edited by Peter Urbach and John Gibson. Chicago: Open Court, 1994.
Francis Bacon: A Critical Edition of the Major Works, ed. Brian Vickers. New York: Oxford University Press, 1996.

Criticism

Anderson, F. H. *The Philosophy of Francis Bacon*. Chicago: University of Chicago Press, 1948.

Briggs, John C. *Francis Bacon and the Rhetoric of Nature.* Cambridge: Harvard University Press, 1989.

Gaukroger, Stephen. *Francis Bacon and the Transformation of Early-Modern Philosophy.* Cambridge: Cambridge University Press, 2001.

Jardine, Lisa. *Francis Bacon: Discovery and the Art of Discourse.* Cambridge: Cambridge University Press, 1974.

Minkov, Svetozar Y. *Francis Bacon's "Inquiry Touching Human Nature": Virtue, Philosophy, and the Relief of Man's Estate.* New York: Lexington, 2010.

Price, Bronwen, ed., *Francis Bacon's New Atlantis: New Interdisciplinary Essays.* New York: Manchester University Press, 2002.

Quinton, Anthony. *Francis Bacon.* Oxford: Oxford University Press, 1980.

Rees, Graham. "Bacon's Speculative Philosophy," in *The Cambridge Companion to Bacon,* ed. Markku Peltonen. Cambridge: Cambridge University Press, 1996, pp. 121–45.

Rossi, Paolo. *Francis Bacon: From Magic to Science.* Translated by Sacha Rabinovitch. Chicago: University of Chicago Press, 1968.

Solomon, Julie Robin and Catherine Gimelli Martin, eds. *Francis Bacon and the Refiguring of Early Modern Thought: Essays to Commemorate The Advancement of Learning (1605–2005).* Burlington: Ashgate, 2005.

Van Malssen, Tom. *The Political Philosophy of Francis Bacon: On the Unity of Knowledge.* Albany: State University of New York Press, 2015.

Vickers, Brian, ed., *Essential Articles for the Study of Francis Bacon.* Hamden, Ct.: Shoe String Press, 1968.

Vickers, Brian. *Francis Bacon and Renaissance Prose.* Cambridge: Cambridge University Press, 1968.

Weinberger, Jerry. "Science and Rule in Bacon's Utopia: An Introduction to the Reading of the *New Atlantis,*" *American Political Science Review* 70 (September 1976): 865–85.

Weinberger, Jerry. *Science, Faith, and Politics: Francis Bacon and the Utopian Roots of the Modern Age.* Ithaca: Cornell University Press, 1985.

Weinberger, Jerry, ed., *Francis Bacon: The History of the Reign of King Henry the Seventh: A New Edition With Introduction, Annotation, and Interpretive Essay.* Ithaca: Cornell University Press, 1996.

White, Howard B. *Peace Among the Willows: The Political Philosophy of Francis Bacon.* The Hague: Martinus Nijhoff, 1968.

Whitney, Charles. *Francis Bacon and Modernity.* New Haven: Yale University Press, 1986.